CHEMICAL OXIDATION

TECHNOLOGIES FOR THE NINETIES

Edited by

W. Wesley Eckenfelder

Eckenfelder, Inc.

Alan R. Bowers

Vanderbilt University

John A. Roth

Vanderbilt University

PROCEEDINGS OF THE FIRST INTERNATIONAL SYMPOSIUM
CHEMICAL OXIDATION: TECHNOLOGY FOR THE NINETIES
VANDERBILT UNIVERSITY
NASHVILLE, TENNESSEE
FEBRUARY 20–22, 1991

TECHNOMIC
PUBLISHING CO., INC.
Lᴀɴᴄᴀsᴛᴇʀ • Bᴀsᴇʟ

Chemical Oxidation
a TECHNOMIC ℠ publication

Published in the Western Hemisphere by
Technomic Publishing Company, Inc.
851 New Holland Avenue
Box 3535
Lancaster, Pennsylvania 17604 U.S.A.

Distributed in the Rest of the World by
Technomic Publishing AG

Printed in the United States of America
10 9 8 7 6 5 4 3 2

Main entry under title:
  Chemical Oxidation: Technologies for the Nineties

A Technomic Publishing Company book
Bibliography: p.

Library of Congress Card No. 91-66929
ISBN No. 87762-895-5

HOW TO ORDER THIS BOOK

BY PHONE: 800-233-9936 or 717-291-5609, 8AM–5PM Eastern Time
BY FAX: 717-295-4538
BY MAIL: Order Department
Technomic Publishing Company, Inc.
851 New Holland Avenue, Box 3535
Lancaster, PA 17604, U.S.A.
BY CREDIT CARD: American Express, VISA, MasterCard

TABLE OF CONTENTS

PREFACE

New regulations governing the discharge of toxic pollutants has focused increased emphasis on physical-chemical technologies which can treat these pollutants in a cost effective manner. Foremost, among these technologies is chemical oxidation.

While chemical oxidation technology has been known and used for many years, its application to wastewater treatment is relatively recent. This volume, the first of its kind, focuses on present, state of the art chemical oxidation technologies with regard to various wastewater applications. The chemical principles intrinsic to the most common oxidants are discussed, including chemical analysis of oxidants as well as the theory of oxidation and reactor design. Specific industrial applications based on current research and case studies are presented. In addition, the use of emerging technology alternatives to specific applications is developed.

This volume should prove a valuable aid to engineers and scientists engaged in developing cost effective solutions to complex water quality problems in today's regulatory environment.

W. W. Eckenfelder
A. R. Bowers
J. A. Roth

The Role of Chemical Oxidation in Wastewater Treatment Processes

ABSTRACT

Chemical oxidation can be applied to the oxidation of chemicals in groundwaters, pretreatment of hazardous and toxic wastewaters and in some cases post treatment for detoxification. This paper presents examples of these applications.

A protocol for screening chemical oxidation applications is described and several examples are presented.

Chemical oxidation can be applied for one of several purposes in a wastewater treatment process depending on the results desired. These are:

- treatment of toxic organics at low concentrations in groundwater

- treatment of low volume-high strength wastewaters for detoxification and enhanced biodegradability

- treatment of wastewaters not normally subject to bio-oxidation, e.g., cyanide and complexed metals

- detoxification relative to aquatic toxicity following biological treatment

Normally, some form of biological treatment will be the most cost-effective technology for organic wastewaters. Some organics, however, are either non-biodegradable or are toxic either to the biological process or to aquatic life and must therefore be pre or post-treated. Chemical oxidation is one of the candidate processes for this purpose. Alternative applications of chemical oxidation are shown in Figure 1.

Many contaminated groundwaters contain a variety of organics many of which can be chemically oxidized. The technology of choice will usually be catalyzed hydrogen peroxide or catalyzed on non-catalyzed ozone. The technology can be applied in-situ or as a surface treatment of the groundwater. At this time, I know of no full-scale applications of in-situ groundwater treatment. Results of the treatment of one groundwater with catalyzed

W. Wesley Eckenfelder, Jr., Eckenfelder Inc., Nashville, Tennessee, U.S.A.

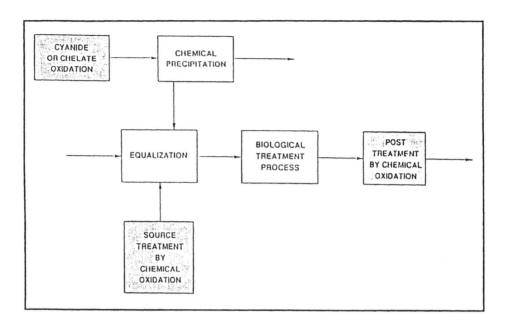

Figure 1. Alternative Applications of Chemical Oxidation

Dosage (mg/l)	H2O2 pH (su)	Ace + IPA (mg/l)	MEK (mg/l)	2-But (mg/l)	4M-2P (mg/l)	MIBK (mg/l)	THF (mg/l)
					Reactor Concentration[a]		
0	5.2	2,860	304	1,010	1,410	792	55
57	5.2	2,790	300	918	1,310	770	41
57	3.5	6,510	342	350	657	127	31
113	3.1	3,754	14	ND	ND	ND	ND

TABLE I - EFFECT OF HYDROGEN PEROXIDE WITH FERROUS ION CATALYST ON REMOVAL OF SEMI-VOLATILE ORGANICS

[a]One hour reaction time.
[b]Adjusted with H_2SO_4.

hydrogen peroxide is shown in Table I. Depending on the constituents in the groundwater, chemical oxidation may be used in conjunction with other technologies. These are stripping of volatile organics, granular carbon adsorption or biological oxidation through biologically activated carbon columns or other processes depending on the concentration of organics in the groundwater. A feasibility study should be conducted to determine the most cost-effective sequence of processes.

Some organics and process wastewaters are inhibitory to biological treatment or are toxic to aquatic life following biological treatment. In many cases, pretreatment by chemical oxidation will effect detoxification as well as enhanced biodegradation. An example of this for several chlorinated organics is shown in Table II. Several organics oxidizable with catalyzed hydrogen peroxide is shown in Table III and those oxidizable with ozone shown in Table IV.

TABLE II - HYDROGEN PEROXIDE OXIDATION OF ORGANICS

Compound (mg/l)	Initial Conc.	Percent Reduction COD	TOC	LC$_{50}$ (%) Before Oxidation	After Oxidation	COD Reduction in 2 days (%) Before Oxidation	After Oxidation
Nitrobenzene	616	72	38	6.0	76.2	59	31
Aniline	466	77	43	35.7	NT	0	40
o-Cresol	541	75	56	2.5	NT	16	51
m-Cresol	541	73	38	1.3	NT	0	51
p-Cresol	541	72	40	0.4	NT	65	47
o-Chlorophenol	625	75	48	5.1	NT	18	37
m-Chlorophenol	625	75	41	1.8	NT	0	39
p-Chlorophenol	625	76	22	0.3	NT	0	39
2,3-DCP	8.5	70	53	1.0	NT	`12	31
2,4-DCP	815	69	50	0.6	NT	9	32
2,5-DCP	815	74	42	1.9	NT	14	38
2,6-DCP	815	61	33	5.7	17.3	0	9
3,5-DCP	815	69	49	0.5	NT	0	9
2,3-DNP	921	80	51	6.3	85.6	0	19
2,4-DNP	921	73	51	2.0	NT	0	49
2,4,6-TCP	800	47	44	2.8	52.2	0	39

Conditions--stoichiometric dosage of H_2O_2, pH 3.5, 50 mg/l Fe^{++}.
NT = Not Toxic

TABLE III - ORGANICS REMOVED BY HYDROGEN PEROXIDE WITH UV CATALYST

Trichloroethylene
Tetrachloroethylene
2-Butanol
Chloroform
Methyl isobutyl ketone
4-Methyl-2-pentanol
Methyl ethyl ketone
Carbon tetrachloride
Tetrachloroethylene

3

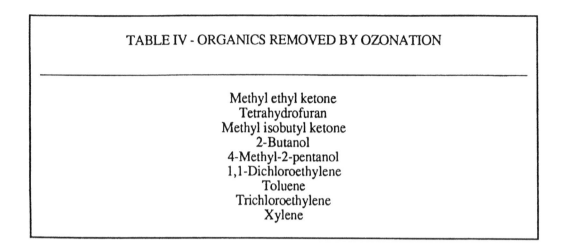

TABLE IV - ORGANICS REMOVED BY OZONATION

Methyl ethyl ketone
Tetrahydrofuran
Methyl isobutyl ketone
2-Butanol
4-Methyl-2-pentanol
1,1-Dichloroethylene
Toluene
Trichloroethylene
Xylene

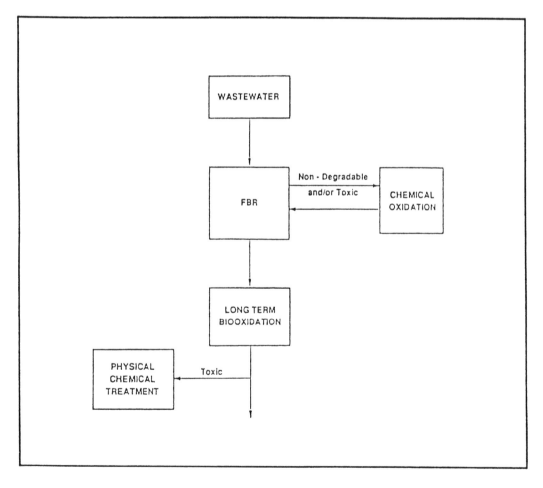

Figure 2. Protocol for Evaluation of Chemical Oxidation for Detoxification and Enhanced Biodegradability

If we are dealing with known specific organics and there is data available on the efficacy of chemical oxidation, studies need only be conducted on the toxicity or biodegradability of oxidation by-products, if any. In this case, the protocol shown in Figure 2 is employed following chemical oxidation of the wastewater. Both the biodegradability and residual aquatic toxicity after biooxidation should be evaluated.

In cases of an unknown wastewater, toxicity to the biological process should first be evaluated using the fed batch reactor (FBR) technique or some other evaluation of toxicity. The FBR is shown in Figure 3. In the FBR test, wastewater is added at a constant rate to activated sludge in a reactor and the rate of biodegradation observed over time. The rate of biodegradation is computed as the difference between the slope with no degradation (Curve A) and the slope computed from the residual contaminant level in the reactor (Slope B). Toxicity is identified by a change in the biodegradation rate or decrease in the oxygen uptake rate. The detailed procedure is described by Eckenfelder (1). If the wastewater is non-degradable and exhibits toxicity, a chemical oxidation study should be conducted.

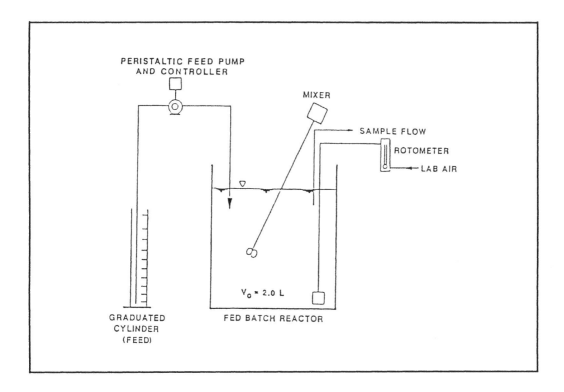

Figure 3. Fed Batch Reactor Configuration

The FBR procedure is then repeated on the wastewater after chemical oxidation in order to determine reduced toxicity and enhanced biodegradability. Results for one wastewater using the FBR test before and after chemical oxidation with permanganate is shown in Figure 4.

5

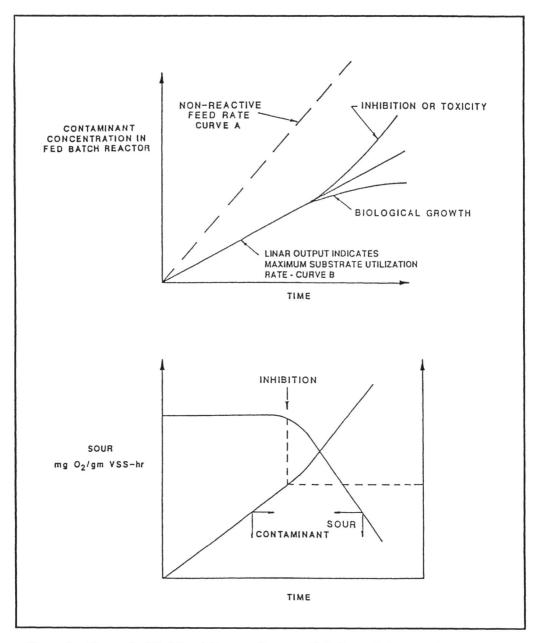

Figure 3a. Theoretical Fed Batch Reactor Output with Influent Substrate Mass Flow Rate Greater than $q_{max} \cdot X_V$ and Inhibition Effects

In order to determine the effect of biological treatment on effluent toxicity, a long-term biooxidation study should be conducted. If acclimated sludge is available, a 48 hour aeration at an organic loading (F/M) of 0.2 should result in removal of all the degradable organic matter. If the sludge is not acclimated, a long-term oxidation test of 14 days to 28 days such as the Zahn-Wallens procedure should be employed (2). Aquatic toxicity on the effluent should again be evaluated after the long-term oxidation study.

The application of chemical oxidation to the detoxification of a specific industrial wastewater stream is shown in Figure 5. A non-degradable component of the wastewater, as evidenced by an increase in effluent COD, inhibited biooxidation as shown by an increase in effluent BOD. Detoxification through treatment with hydrogen peroxide is shown in Figure 6. The volume of the toxic wastewater is low so that even with high chemical dosages the treatment is cost effective.

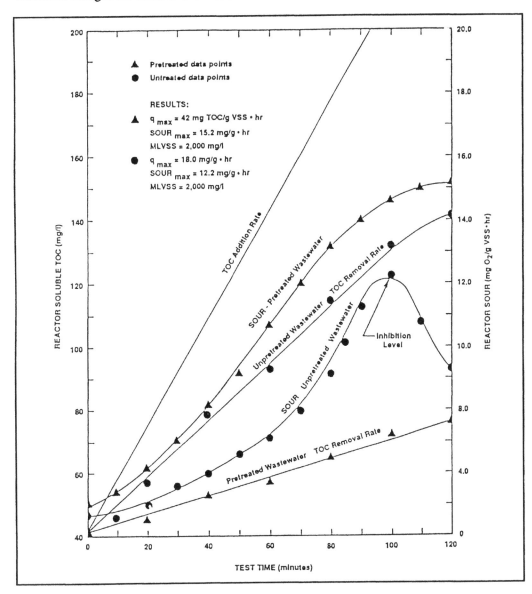

Figure 4. Chemical Oxidation of an Organic Chemicals Wastewater with Permanganate to Enhance Biodegradability

Wastewater may contain constituents which are not amenable to biological treatment or which inhibit or prevent other processes from being effective. One example is cyanide, which can be effectively oxidized using chlorine or hydrogen peroxide. When heavy metals are organically chelated, they cannot be removed by conventional chemical precipitation. In many cases, such as EDTA, chemical oxidation will oxidize the chelation thus freeing the metal for preciptiation. $[Pb \cdot EDTA]^{-3}$ will oxidize with ozone at a Ph of 9 in 15 minutes to yield PbO_2. Similar reactions occur with other metals.

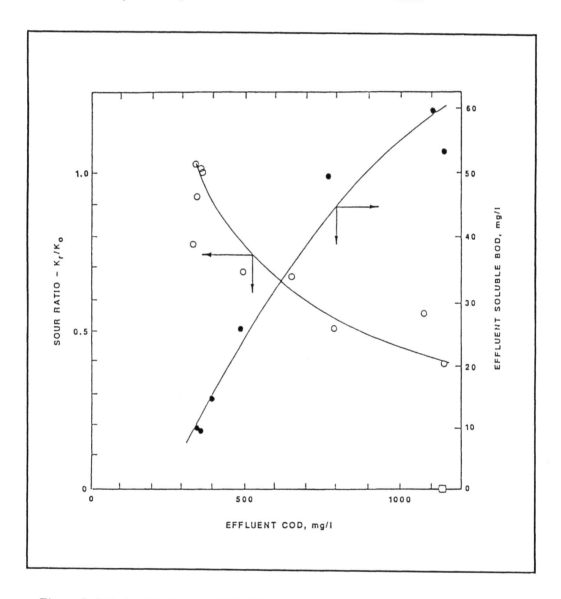

Figure 5. Relationship Between Uninhibited and Inhibited SOUR and the Effluent BOD

In many cases, an effluent after biological oxidation is still toxic due to the presence of non-degradable constituents or to biooxidation by-products. In some cases, the COD or TOC is in excess of permit requirements. Chemical oxidation may reduce the toxicity and the COD. One possible problem is to generate more degradable by-products, thus increasing the effluent BOD. This application of chemical oxidation is probably limited by the economics involved, but should find application in specific cases.

In summary, chemical oxidation is a competitive process to carbon adsorption or air or steam stripping of volatile organics. The economics of the process should improve with developments of enhanced catalysis and in the case of ozone improved mass transfer. A recent reported development is the generation of microbubbles of ozone which has greatly enhanced the efficiency of ozone oxidation.

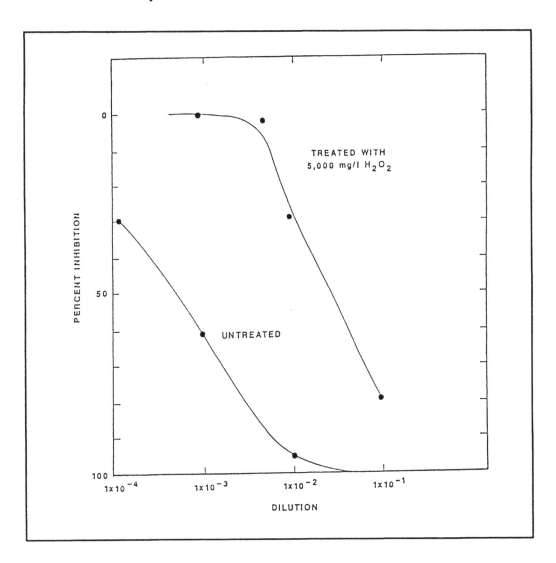

Figure 6. Detoxification of an Industrial Wastewater

REFERENCES

1. Eckenfelder, W. W. , *Industrial Water Pollution Control,* 2nd Ed. McGraw-Hill Book Company, New York, NY, p 182.

2. Pitter, P. and Chudoba, J., *Biodegradability of Organic Substances in the Aquatic Environment*, CRC Press Inc., Boca Ratan, 1990, p. 146.

A. R. BOWERS
S. H. CHO
A. SINGH

Chemical Oxidation of Aromatic Compounds: Comparison of H_2O_2, $KMnO_4$ and O_3 for Toxicity Reduction and Improvements in Biodegradability

ABSTRACT

A variety of aromatic organics (15 in all) were partially oxidized using hydrogen peroxide, ozone , or potassium permanganate. Of the compounds tested, all 15 reacted with hydrogen peroxide, 14 of 15 reacted with ozone and 7 of 15 reacted with permanganate. Oxidation with hydrogen peroxide resulted in 17 to 79% ultimate conversion of the organic carbon to carbon dioxide, while ozone resulted in 0 to 74% ultimate conversion and the 7 that reacted with permanganate yielded 22 to 68% ultimate conversion. The toxicity of the reactive compounds was typically greatly reduced (based on Microtox), however several compounds exhibited an increase in toxicity after oxidation (on an equivalent TOC basis); resorcinol, vanillin and salicylic acid for hydrogen peroxide; resorcinol, and sulfanilic acid for permanganate. Using an unacclimated municipal activated sludge, the oxidized compounds were found to be much more amenable to biological treatment on the basis of overall removal from solution (chemical oxidation + bio-oxidation vs bio-oxidation only).

INTRODUCTION

A variety of organic compounds are refractory, toxic or inhibitory in nature. Wastewaters containing these organics must be pretreated before being introduced to conventional biological waste treatment systems. The purpose of pretreatment is to remove or destroy these compounds, or alternately to convert them to less toxic and more readily biodegradable intermediate products which can be more easily treated by conventional biological methods. One alternative is pre-oxidation of these compounds using chemical oxidants, such as hydrogen peroxide, permanganate or ozone [1,2].

A. R. Bowers, Associate Professor, Dept. of Civil and Environmental Engineering, Vanderbilt University, Nashville, Tennessee, U.S.A.
S. H. Cho, Associate Professor, Dept. of Environmental Engineering, Ajou University, Suwon, Korea
A. Singh, Advent, Inc., Brentwood, Tennessee, U.S.A.

Hydrogen peroxide, combined with a ferrous iron catalyst, known as Fenton's reagent, has been shown effective in treating a variety of phenolic compounds [3,4]. Studies using solutions of pure compounds and industrial wastewaters indicated near complete destruction of the original compounds during oxidation [3-6]. In addition, the oxidation by-products were less toxic and more biodegradable than the original phenolic compounds.

Permanganate is another vigorous oxidant which has shown promise. Long known to be effective in removing tastes and odors caused by the decomposition of organic matter, permanganate has demonstrated effectiveness in destroying phenolic compounds. The reactivity of permanganate depends on the types of organic compounds present and to a great extent on the reaction pH conditions [2,7].

Ozone is also known to be a powerful oxidant available for practical application in water and wastewater treatment. It has been used for disinfection, chemical oxidation and pre-oxidation in preparation for biological treatment. However, ozone is considered to be much more selective in its reactions than hydrogen peroxide and permanganate [8-10]. The general factors governing the treatment efficiency is the selectivity of ozone for the target compound, pH, extent of incidental stripping associated with ozone mass transfer dispersion, and temperature [2,8-10].

This study was undertaken to compare the reactivity of these oxidants with a variety of aromatic organic compounds. Using typical reaction conditions, the extent of destruction of the original compounds and characteristics of the reaction by-products were examined, i.e., ultimate conversion to CO_2, toxicity and biodegradability.

MATERIALS AND METHODS

TEST COMPOUNDS

All compounds were obtained in the purest forms commercially available (99% or better) from Fisher Scientific, Inc., Aldrich Chemical, Inc., or Fluka Chemical Company. Stock solutions were prepared based on the solubility of the organic compounds, i.e., 0.1M unless this exceeded the solubility.

OXIDANTS AND REACTION CONDITIONS

Reaction conditions for these tests were selected to be consistent with known reactivities for each oxidant, i.e., pH and catalyst. At the end of each reaction period, any residual oxidant was destroyed to prevent later interference with sample analyses. A summary of the reaction conditions and reducing agents for destruction of residual oxidants is reported in Table I.

The oxidant dosages were based on the required stoichiometric amount for each organic compound. This was calculated assuming complete or ultimate oxidation of each individual compound to CO_2, or using catechol, $C_6H_4(OH)_2$, as an example:

H_2O_2:
$$C_6H_4(OH)_2 + 13H_2O_2 \rightarrow 6CO_2 + 16H_2O \tag{1}$$
$KMnO_4$:
$$C_6H_4(OH)_2 + 26/3MnO_4^- + 26/3H^+ \rightarrow 26/3MnO_2 + 6CO_2 + 22/3H_2O \tag{2}$$
O_3:
$$C_6H_4(OH)_2 + 13O_3 \rightarrow 6CO_2 + 3H_2O \tag{3}$$

CHEMICAL OXIDATION PROCEDURE

Oxidation reactions were carried out in batch shake flask tests in which the oxidant, catalyst and organic compounds were brought together and the initial pH adjusted to 3.5. For hydrogen peroxide and permanganate a 24-hour reaction time was provided. For ozone, which had to be input as a gas, mass transfer and self-decomposition had to be accommodated. Therefore, approximately 10x the stoichiometric amount was infused to the reactors (based on flow rates and gaseous O_3 composition measurements). It was noted that no residual ozone existed after 4 hours of reaction, so 4 hours was used as the reaction period.

Table I. Reaction Conditions and Method of Destruction of Residual Oxidants

Oxidant	Reaction pH	Catalyst	Residual Reducing Agent
H_2O_2[a]	3.5[b]	$FeSO_4$[c]	Fermcolase-1000[d]
$KMnO_4$[e]	3.5[f]	none	Na_2SO_3[e,g]
O_3[h]	3.5[i]	none	none[j]

a: Obtained from Interox America, Inc., as 50% by weight.
b: Based on previous data for free radical generation and reaction (3-6)
c: 20 mg/L (as Fe).
d: Finnsugar Biochemicals, Inc., Schaumburg IL; absence of residual H_2O_2 was verified using H_2O_2 test strips (EM Science, Inc., Cherry Hill, NJ).
e: Fisher Scientific, Inc.
f: For pH 3.5 to 7.0; $MnO_4^- + 4H^+ + 3e^- \rightarrow MnO_2 + 2H_2O$ (2)
g: Precipitated MnO_2 was removed using prewashed 0.45 μ filters (Gelman No. GN-6).
h: Using a laboratory ozone generator (LINDE Model SG -4060).
i: O_3 as reactant rather than free radicals (2,10).
j: No O_3 residual was ever detected.

BIO-OXIDATION PROCEDURE

Biodegradability was measured using municipal activated sludge obtained from the Metro Nashville Central Wastewater Treatment Facility. A 14-day batch test was performed with the unoxidized compound and oxidized compound adjusted to the same initial concentration, 100 mg/l (as COD) to prevent inhibitory conditions. The pH of the test solution was maintained at 7.0 using a phosphate buffer. This phosphate buffer also served as the nutrient source for bacterial growth. In addition to this, nitrogen was added. The other trace minerals were supplied in sufficient quantity by adding tap water as the dilution water.

ANALYTICAL TECHNIQUES

COD, TOC and phenolic residual concentration were measured to monitor the decomposition of the organic compound by chemical and/or biological oxidation. For the COD test, the closed reflux, titrimetric method was used. The residual phenolic compounds were measured by the 4-amino antipyrine direct photometric method. For TOC analysis of samples, a Dohrman (Model DC 90) total organic carbon analyzer was used. All of the analytical procedures were conducted in accordance with the methods prescribed by Standard Methods [11].

Toxicity was measured using the Microtox^R (Beckman Model 2055) toxicity analysis. Toxicity of each organic, expressed as an EC50 was measured before and after chemical oxidation, to compare the degree of toxicity using the different oxidants. For these tests, 15 minutes exposure time and 15°C were selected as operating parameters procedures are described in the Microtox System Operating Manual [12].

EXPERIMENTAL ORGANIZATION

Synthetic waste samples were split into two portions, one for bio-oxidation only and one for chemical pre-oxidation followed by bio-oxidation. A flow chart of the experimental plan is given in Figure 1.

RESULTS AND DISCUSSION

Fifteen aromatic compounds were reacted with hydrogen peroxide, ozone, or permanganate under the conditions outlined previously. For these compounds that reacted, almost no residual of the original compound was detected (typically about 1% or less). Both COD and TOC were used to evaluate the extent of reaction, since COD measures the change in the parent structure and the TOC determines the fraction converted to CO_2. In addition, the ratio of COD to TOC

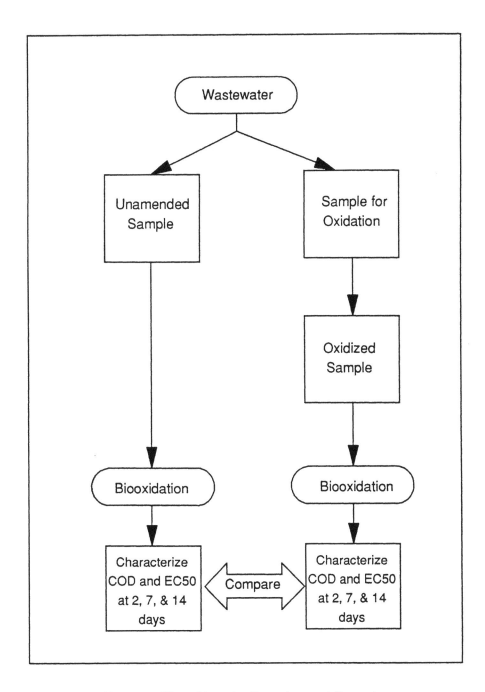

Figure 1. Flow Chart for Experimental Procedure

can be used as a measure of the average oxidation state of the residual organic carbon, or [13]:

$$\text{Average Oxidation State} = 4(TOC-COD)/TOC \qquad (1)$$

where, TOC and COD are in Molar units.

REACTIVITY

The results of the chemical oxidation tests for the 15 compounds using H_2O_2, $KMnO_4$ and O_3 are presented in Table II. A comparison of the results indicates that H_2O_2 was reactive with all 15 organic compounds, while $KMnO_4$ reacted with only 8 out of 15 and O_3 reacted with 14 out of 15. For the compounds that reacted, similar ultimate degradation (TOC removal) and COD removals were obtained, although the COD removal was significantly less on the average for O_3 compared to H_2O_2 and $KMnO_4$. Mostly, these results were expected. Hydrogen peroxide reacts as the result of the production of free radicals that exhibit little selectivity and thus react with a wide variety of organics (all of them in this case) while O_3 is the reactive species in the lower pH region exhibiting some degree of selectivity (nonreactive with coumarin in this case) that contributes to the lesser extent of reaction (only 55.9% COD removal on the average), and MnO_4^- would be expected to exhibit some selectivity as well (7 non-reactive compounds in this case). While this appears to be a disadvantage, the selectivity of permanganate may be an advantage where undesirable organics are swamped by a mixture of compounds that are biodegradable and non-toxic. Additionally, other conditions (pH < 3.5 or pH > 10), may have yielded a larger number of reactive compounds.

For those compounds that reacted, a summary of the mean oxidation states of the organic carbon before and after oxidation is shown in Figure 2. In general, the original organic carbon was converted to more highly oxidized organic ecompounds and 3 regions exist for the oxidation products. For H_2O_2, all but one compound (see H5) were converted to oxidation states greater than +1.6, while most of the $KMnO_4$ products ranged from +1.2 to +2.3 (see M11 and M7 as exceptions), and O_3 was the least aggressive, yielding residual carbon in the -0.3 to +1.6 oxidation states.

TOXICITY

Previous work using H_2O_2 as the only oxidant showed that the toxicity of the oxidation by-products was reduced for all of the compounds tested [1-5]. For the compounds tested in this study, the toxicity of the original compounds and the toxicity of the organic by-products after oxidation with HO_2, $KMnO_4$, and O_3 are presented for comparison in Table III.

Table II. Chemical Oxidation of Various Aromatic Compounds
Using H_2O_2, $KMnO_4$, and O_3

Compound[a]	Initial Oxidation State[b]	TOC Removal (%)[c,d,e]			COD Removal (%)[d,e]		
		H_2O_2	$KMnO_4$	O_3	H_2O_2	$KMnO_4$	O_3
1. Pyrrolidine	-1.76	34.9	NR	32.1	72.1	NR	58.5
2. Sulfanilic Acid	-0.84	46.3	NR	57.5	74.9	NR	57.4
3. Naphthalene	-0.80	46.2	NR	0.0	80.4	NR	>99.0
4. Diphenylamine	-0.66	69.4	NR	30.6	87.7	NR	90.0
5. Skatole	-0.66	0.0	NR	0.0	39.0	NR	38.1
6. Benzaldehyde	-0.57	78.6	67.6	74.4	93.5	79.1	74.2
7. Indole	-0.50	62.3	60.3	60.9	95.5	91.0	77.2
8. Catechol	-0.33	57.0	52.2	22.0	80.5	66.3	30.7
9. Hydroquinone	-0.33	30.7	27.3	17.2	78.5	71.2	45.0
10. Resorcinol	-0.33	56.5	27.8	29.1	79.8	73.1	50.1
11. Vanillin	-0.25	70.3	53.4	63.6	87.8	55.2	63.6
12. Pyrogallol	0.00	45.4	22.1	28.5	75.1	78.2	48.5
13. Salicylic Acid	0.00	28.6	31.6	31.2	74.6	49.8	41.6
14. Coumarin	+0.22	25.9	NR	NR	65.3	NR	NR
15. Phthalic Acid	+0.25	37.0	NR	31.1	71.2	NR	52.0
Average[f]	-0.44	45.9	42.8	36.8	77.1	70.5	55.9

a: Arranged in order of lowest to highest oxidation state of organic carbon.
b: Mean oxidation state of the organic carbon calculated from molecular structure.
c: Corresponds to the percentage converted to CO_2, i.e., ultimate oxidation.
d: All compounds were initially at a concentration of 5×10^{-3} M (as the original compound) unless limited by solubility (Naphthalene and Diphenylamine).
e: NR - non reactive (no measurable reduction in COD).
f: Average based on reactive compounds only.

Table III. Toxicity of Compounds Before and After Oxidation

Compound	Original EC50[a] (%)	EC50 After Chemical Oxidation(%)[a,b]		
		H_2O_2	$KMnO_4$	O_3
1.Pyrrolidine	NT[c]	NT	NR[d]	NT
2.Sulfanilic Acid	14.0(49.6)	62.0(117.8)	NR	70.0(107.1)
3.Naphthalene	41.0(5.3)	NT(7.0)[d]	NR	NT(14.0)[e]
4.Diphenylamine	14.0(5.0)	83.0(9.1)	NR	88.0(22.0)
5.Skatole	0.9(2.6)	4.2(12.6)	NR	8.0(24.2)
6.Benzaldehyde	4.8(20.2)	54.0(48.6)	11.0(15.0)	39.0(42.1)
7.Indole	1.3(6.2)	56.0(100.8)	25.0(47.5)	45.0(84.6)
8.Catechol	8.1(28.4)	30.0(45.3)	49.5(95.5)	39.0(108.0)
9.Hydroquinone	0.3(1.0)	1.6(3.9)	63.0(172.0)	48.0(143.0)
10.Resorcinol	55.0(194.7)	64.0(98.6)	42.0(107.9)	48.0(120.5)
11.Vanillin	22.0(103.8)	53.0(74.2)	52.0(236.4)	65.0(113.1)
12.Pyrogallol	2.8(9.4)	61.0(111.6)	50.0(130.5)	42.0(102.1)
13.Salicylic Acid	27.0(110.4)	25.0(73.0)	26.0(90.2)	28.0(98.3)
14.Coumarin	2.0(12.4)	NT(460.0)[d]	NR	NR
15.Phthalic Acid	NT	NT	NR	NT

a: Based on Microtox Toxicity; % volume (or concentration) to reduce the light produced from luminescence by 50%.
b: Values in parenthesis represent the EC50 on a TOC basis (as mg/L).
c: NT = non-toxic, i.e., no detectable light reduction.
d: NR = nonreactive compound for this particular oxidant.
e: Value in parenthesis for non-toxic compounds represents the highest TOC value at no dilution.

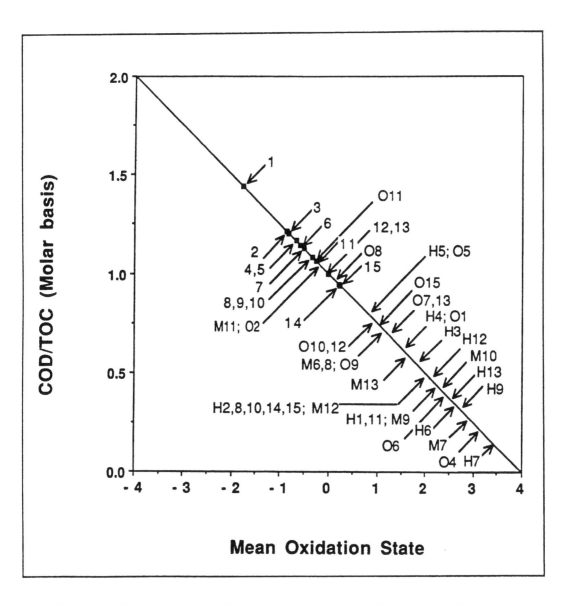

Figure 2. Mean Carbon Oxidation State Before and After Chemical Oxidation
Note: Data points indicate original oxidation states while
prefixes H, M, and O refer to oxidation products from
H_2O, $KMnO_4$, and O_3, respectively.

In general, the oxidation products consisted of compounds that were less toxic than the original compounds. For H_2O_2, 10 of the compounds that were initially toxic were improved after oxidation (on a TOC basis), i.e, improvements ranged from 1.6 to > 37.1x less toxic. However, for three compounds the toxicity increased, i.e, resorcinol, vanillin and salicylic acid (2.0, 1.4 and 1.5x more toxic, respectively). For $KMnO_4$, 5 out of 8 reactive compounds decreased in toxicity (1.3 to 172x less toxic) while three compounds increased in toxicity, i.e, benzaldehyde, resorcinol and salicylic acid (1.3, 1.8, and 1.2x more toxic, respectively). For O_3, 12 out of 14 reactive compounds exhibited a decrease in toxicity (1.1 to 143x less toxic) while 2 compounds showed an increase in toxicity, i.e., resorcinol and salicylic acid (1.1 and 1.6x more toxic, respectively). In common, the oxidation by-products for resorcinol and salicylic acid showed a slightly elevated toxicity for all three oxidants (1.1 to 2.0x more toxic). The apparent order of toxicity among the original and oxidized compounds is summarized as follows:

Before Oxidation

Original Compounds:

9 > 5 > 4 > 3 > 7 > 12 > 14 > 6 > 8 > 27 > 11 > 13 > 10

After Oxidation

H_2O_2:

9 > 4 > 5 > 8 > 6 > 13 > 11 > 10 > 7 > 12 > 2 > 3 > > 14

$KMnO_4$:

6 > 7 > 13 > 8 > 10 > 12 > 11 > 9

O_3:

4 > 5 > 6 > 7 > 13 > 12 > 1 > 8 > 11 > 10 > 9

It is interesting to note that hydroquinone exhibited the greatest toxicity of the original compounds and the H_2O_2 oxidized compounds, but the least toxicity of the $KMnO_4$ and O_3 oxidized compounds. However, most of the other compounds are in similar order with little difference in the magnitudes of toxicity.

Biodegradability

The biodegradabilities of the original compounds and oxidation products were evaluated using unacclimated municipal activated sludge under aerobic conditions. This permitted an evaluation of how readily biodegradable the compounds were compared to those typically encountered in a municipal activated sludge system. The reduction in COD at 2 and 14-days is presented in Table IV for the original and oxidized compounds.

At first comparison, it appears that chemical oxidation retards the biodegradability of many of the compounds. For example, after 14 days, oxidation products of compounds 1, 6-11, 14 and 15 exhibit significantly less COD reduction during biodegradation than the original compounds. However, these compounds have already undergone substantial chemical oxidation and the "apparent" difference in residual COD is an artifact of the dilution procedure prior to biodegradation (100 mg/L initial COD). Therefore, a correction procedure was applied based on the overall removal as follows:

$$f_{residual} - [\frac{COD_{ox}}{COD_o}][\frac{COD_{br}}{100}] \qquad (2)$$

where $f_{residual}$ = overall fraction of residual COD after chemical and bio-oxidation

COD_o = initial COD of waste before chemical oxidation

COD_{ox} = initial COD of waste before chemical oxidation

COD_{br} = residual COD after bio-oxidation based on an original COD of about 100 mg/L by
                   dilution

These values were calculated for each set of oxidation products and are shown as values in parenthesis in Table IV. On this basis, chemical oxidation followed by bio-oxidation was superior to bio-oxidation only in most cases. The most notable exceptions include pyrrolidine (roughly equivalent in all comparisons), skatole (H_2O_2 and O_3 result in 3x the residual COD), catechol ($KMnO_4$ and O_3 yield 2 to 4x more residuals), salicylic acid ($\sim$ 2 to 5x more residuals), and coumarin (H_2O_2 yielded 2x more residuals).

SUMMARY

Hydrogen peroxide, $KMnO_4$, and O_3 were composed as oxidants for a variety of aromatic organic compounds. Of the three oxidants, H_2O_2 was most universally reactive, 15 out of 15 compounds, followed closely by ozone, 14 out of 15, and then permanganate, 8 out of 15. In most cases, the toxicity of the oxidation products on a TOC basis, was reduced compared to the original compound. In addition, the residual organics, in terms of COD, were less after chemical + bio-oxidation than

Table IV. Residual COD of Compounds Undergoing Biodegradation

Compound	2 Days[a]				14 Days[a,b]			
	Unoxidized	H₂O₂	KMnO₄	O₃	Unoxidized	H₂O₂[c]	KMnO₄[c]	O₃[c]
1.Pyrrolidine	0.17	0.47	-	0.45	0.11	0.37 (0.10)	-	0.38 (0.15)
2.Sulfanilic Acid	0.48	0.18	-	0.22	0.42	0.16 (0.04)	-	0.18 (0.07)
3.Naphthalene	0.04	0.22	-	-	0.00	0.00	-	0.00
4.Diphenylamine	0.99	0.96	-	0.87	0.42	0.18 (0.02)	-	0.18 (0.02)
5.Skatole	0.28	0.49	-	0.50	0.09	0.42 (0.26)	-	0.44 (0.27)
6.Benzaldehyde	0.33	0.94	0.73	0.93	0.21	0.79 (0.05)	0.33 (0.07)	0.40 (0.10)
7.Indole	0.23	1.00[d]	1.00[d]	1.00[d]	0.16	1.00[d] (0.05)[e]	1.00[d] (0.09)[e]	1.00[d] (0.23)
8.Catechol	0.18	0.69	0.52	0.49	0.06	0.37 (0.07)	0.33 (0.11)	0.38 (0.26)
9.Hydroquinone	0.98	0.52	0.28	0.49	0.29	0.30 (0.06)	0.23 (0.07)	0.28 (0.15)
10.Resorcinol	0.19	0.47	0.56	0.36	0.11	0.31 (0.06)	0.22 (0.06)	0.30 (0.15)
11.Vanillin	0.16	0.29	0.14	0.33	0.09	0.22 (0.03)	0.09 (0.04)	0.20 (0.07)
12.Pyrogallol	0.49	0.59	0.51	0.54	0.38	0.38 (0.09)	0.38 (0.08)	0.43 (0.22)
13.Salicylic Acid	0.18	0.44	0.82	0.81	0.05	0.36 (0.09)	0.50 (0.25)	0.33 (0.19)
14.Coumarin	0.12	0.64	-	-	0.09	0.63 (0.18)	-	-
15.Phthalic Acid	0.50	0.77	-	0.73	0.17	0.42 (0.12)	-	0.39 (0.19)

a: Values represent fraction of initial COD based on 100 mg/L COD initially.
b: Values in parenthesis are overall fraction of COD removal.
c: Values in parenthesis are the adjusted overall removal based on Eq. 2.
d: No apparent removal of COD, actual values were greater than initial values.
e: Value in parenthesis represents residual COD after chemical oxidation only.

Table V. Qualitative Summary of Chemical Oxidant Effectiveness in Reducing Toxicity and Improving Overall Removal of Original Organic Compounds

Compound	Reduced Toxicity[a]			Improved Overall Removal[a]		
	H_2O_2	$KMnO_4$	O_3	H_2O_2	$KMnO_4$	O_3
1.Pyrrolidine	na[b]	na	na	Same	NR	No
2.Sulfanilic Acid	Yes	NR[c]	Yes	Yes	NR	Yes
3.Naphthalene	Yes	NR	Yes	-	NR	-
4.Diphenylamine	Yes	NR	Yes	Yes	NR	Yes
5.Skatole	Yes	NR	Yes	No	NR	No
6.Benzaldehyde	Yes	No	Yes	Yes	Yes	Yes
7.Indole	Yes	Yes	Yes	Yes	Yes	No
8.Catechol	Yes	Yes	Yes	Same	No	No
9.Hydroquinone	Yes	Yes	Yes	Yes	Yes	Yes
10.Resorcinol	No	No	No	Yes	Yes	Yes
11.Vanillin	No	Yes	Yes	Yes	Yes	Yes
12.Pyrogallol	Yes	Yes	Yes	Yes	Yes	Yes
13.Salicylic Acid	No	No	No	No	No	No
14.Coumarin	Yes	NR	NR	No	NR	NR
15.Phthalic Acid	na	na	na	Yes	NR	No

a: Compared to original/compound.
b: na = Not applicable - not initially toxic.
c: NR = Compound non-reactive with this oxidant.

bio-oxidation of the original compounds only. A qualitative summary of the results is presented in Table V.

ACKNOWLEDGEMENT

This work was partially supported by Interox America, Inc., Mobile Oil, and Tenneco, Inc.. The support of the Korean Science Foundation (KOSEF) for Soon H. Cho was also greatly appreciated.

REFERENCES

[1] Bowers, A.R., W.W. Eckenfelder, Jr., P. Gaddipati and R.M. Monsen, 1988, "Toxicity Reduction in Industrial Wastewater Discharges," Pollution Engineering, 20, (2), 68.

[2] Bowers, A.R., 1990 "Chemical Oxidation of Toxic, Inhibitory, and Refractory Organics," Chapter 11 in Toxicity Reduction in Industrial Effluents, W. W.Eckenfelder and P.W. Lankford, editors, Van Nostrand Rheinhold, Inc., NY, NY.

[3] Bowers, A.R., W.W. Eckenfelder, Jr., P. Gaddipati, and R.M. Monsen, 1989, "Treatment of Toxic or Refractory Wastewater with Hydrogen Peroxide," Water Science Technol., 21, 477.

[4] Bowers, A.R., W.S. Kong, W.W. Eckenfelder, Jr., and R.M. Monsen, 1989, "Treatment of Aromatic Organic Compounds with Catalyzed Hydrogen Peroxide," Proceedings of the Industrial Waste Symposium, 62nd Annual Conference of the Water Pollution Control Federation, San Francisco, CA.

[5] Kong, W.S., 1989, Masters Thesis, Vanderbilt University, Environmental and Water Resources Engineering, Nashville, TN.

[6] Eisenhaur, H.R., 1964, "Oxidation of Phenolic Wastes, Part I, Oxidation with Hydrogen Peroxide and a Ferrous Salt Reagent," J. Water Pollut. Control Fed., 36, 1116.

[7] Throop, W.M., 1977, "Alternative Methods of Phenol Wastewater Control," J. Hazard. Mater., 1, 319.

[8] Rice, R.G., and A. Netzer, editors, 1982, Handbook of Ozone Technology and Applications, Ann Arbor Science, Inc., Ann Arbor, MI.

[9] Gurol, M.D., 1985, "Factors Controlling the Removal of Organic Pollutants in Ozone Reactors," J. Amer. Water Works Assoc., 77, 77.

[10] Mehta, Y.M., C.E. George, and C.H. Kuos, 1989, "Mass Transfer and Selectivity of Ozone Reactions," Canada J. Chem. Eng., 67, 118.

[11] American Public Health Association, 1985, Standard Methods for the Examination of Water and Wastewater, 16th Edition, APHA - AWWA - WPCF, Washington, DC.

[12] Kasler, L. and M. Greene, 1982, <u>Beckman Microtox System Operating Manual</u>, Belkman Instruments, Carlsbad, CA.

[13] Stumm, W., and J. J. Morgan, 1981, <u>Aquatic Chemistry</u>, 2nd Edition, John Wiley & Sons, Inc., NY, NY.

RANDY A. BULL
JACK D. ZEFF

Hydrogen Peroxide in Advanced Oxidation Processes for Treatment of Industrial Process and Contaminated Groundwater

ABSTRACT

As environmental needs and regulations continue to become more severe through the 1990's, destructive treatment technologies consistent with waste minimization must be developed. Oxidation technologies that can result in on-site destruction of hazardous wastes present a commercially viable alternative to meet these increasingly demanding needs.

Advanced Oxidation Processes (AOP) have the capability of total destruction of many organic pollutants. These systems are particularly applicable to wastewater treatment situations where low discharge limits must be met, and where normally difficult-to-treat chemicals are present. AOP's involving the activation of hydrogen peroxide through combination with ultraviolet light and/or ozone have been successfully implemented as full scale treatment systems in a variety of industrial process water and contaminated groundwater applications.

AOP's are being used as primary water treatment systems to reduce pollutant levels in industrial process wastewater as well as contaminated groundwater. They are also particularly suited for use in combination with other treatment technologies as a pretreatment or as a polish step. In many cases the overall treatment system is more cost-effective when such combinations are implemented. Organic contaminants successfully treated include phenolics, chlorinated solvents, and COD. Commercially operating systems are treating wastewater flows of up to 250 gpm.

In this paper AOP's will be summarized and laboratory and pilot results with specific refractory pollutants will be discussed.

Randy A. Bull, FMC Corporation, Philadelphia, Pennsylvania, U.S.A.
Jack D. Zeff, Ultrox International

Introduction

Regulatory changes and public concern over the environment are requiring waste generators to reduce levels of pollutants in discharged wastewater to extremely low levels. Additionally, to eliminate potential hazards associated with handling and transportation of hazardous wastes, treatment technologies that result in at least partial, if not complete, destruction of pollutants on-site are becoming more desirable. Such on-site destructive waste treatment technologies eliminate the need for transfer and handling of wastes, and thus are consistent with waste minimization strategies being implemented by industry.

Oxidation technologies are capable of achieving desired on-site pollutant destruction. If carried out to its ultimate stage, oxidation can completely oxidize organic compounds to carbon dioxide, water and salts. Partial oxidation can result in increased biodegradability of pollutants so that residual organic compounds can be removed through biological treatment. Many biological processes are, in fact, oxidative processes.

The synthetic organic chemicals that are being regulated to lower levels today are those which do not readily lend themselves to simple oxidative treatment. Depending on the nature of the oxidant, potentially hazardous by-products may be formed, e.g., chlorinated intermediates in chlorination processes. To minimize these by-products, use of oxidants involving oxygen are being investigated. Oxygen itself is generally kinetically too slow to be practical, or requires severe conditions of temperature and/or pressure to effect complete oxidation. Active oxygen products, such as hydrogen peroxide and ozone, are capable of providing the desired oxidation in many cases. However, in the case of refractory, difficult to oxidize organics, these are insufficient in the absence of an activator.

AOP's have been described as oxidation processes that are based on the generation of hydroxyl radical intermediates [1]. AOP's based upon combinations of hydrogen peroxide, ozone and ultraviolet light have been among the most heavily investigated [2-5]. There are a growing number of successful commercial use of these AOP combinations [6-8]. New AOP water treatment processes continue to be discovered, expanding AOP into new treatment applications as they demonstrate effectiveness for the destruction of a broader range of pollutants.

The formation of radical species in AOP is important because of their strong oxidizing potential. A comparison of oxidation potentials is presented in Table I. The hydroxyl radical has an oxidation potential second only to elemental fluorine, making it a desirable intermediate. An oxidant radical is also desirable because of the faster reaction rates (see Table II) compared to molecular oxidants like ozone [9].

Hydrogen Peroxide in Wastewater Treatment

Hydrogen peroxide has long been used in environmental applications, including wastewater treatment, and plays an important role in many AOP's. H_2O_2 is a strong oxidant that continues to find utility in destruction of reduced sulfur compounds, odors, and active

TABLE I. Oxidation Potentials of Radical Species Compared to Common Molecular Oxidants

Oxidant	Potential (volts)
F_2	3.06
$HO^•$	2.80
O_3	2.07
H_2O_2	1.77
$HOO^•$	1.70
$HOCl$	1.49
Cl_2	1.39

TABLE II. Second Order Rate Constants for Ozone & Hydroxyl Radical for a Variety of Compounds

Organic Compound	Rate Constant $(M^{-1}sec^{-1})$	
	O_3 [a]	$HO^•$ [b]
Benzene	2	7.8×10^9
Toluene	14	7.8×10^9
Chlorobenzene	0.75	4×10^9
Trichloroethylene	17	4×10^9
Tetrachloroethylene	<0.1	1.7×10^9
n-butanol	0.6	4.6×10^9
t-butanol	0.03	0.4×10^9

[a]From Hoigne and Bader, 1983 [2].
[b]From Farhataziz and Ross, 1977 [11].

chlorine [10]. When uncatalyzed, H_2O_2 finds applicability primarily for treatment of relatively easy to oxidize inorganic pollutants.

For organic pollutants and compounds like cyanides, H_2O_2 often requires an activator. The most common activators are transition metal salts, for example, iron salts in the classic Fenton's reaction. This system is commonly used to oxidize phenols, as well as other organics. The mechanism involves the catalytic reaction of ferrous ion with H_2O_2 to produce a hydroxyl radical [12]. Because of the useful hydroxyl radical formation, Fenton's reaction is an AOP as currently defined [1]. However, the rate of radical generation is sufficiently slow to make it ineffective for oxidizing certain refractory organics. In addition, the process produces iron sludge which must be disposed.

Hydrogen Peroxide in Advanced Oxidation Processes

To be effective in the treatment of wastewater containing many of today's problem pollutants, a higher rate of radical generation is required. AOP's are capable of producing the necessary radical fluxes for a useful organic destruction process [1]. As a result, these processes are being developed further for application to wastewater treatment. The chemistry of AOP's has been well documented and continues to be the subject of mechanistic and kinetic modeling investigations [13, 14].

Hydrogen peroxide based AOP's offer several approaches:

H_2O_2/ultraviolet light
H_2O_2/ozone
H_2O_2/ozone/ultraviolet light

The chemistry of these combinations can be complex, involving numerous possible AOP reactions as follows [15]:

$$O_3 + H_2O \xrightarrow{h\nu} H_2O_2 + O_2 \tag{1}$$

$$H_2O_2 \xrightarrow{h\nu} 2\,HO^{\bullet} \tag{2}$$

$$H_2O_2 \longrightarrow HO_2^- + H^+ \tag{3}$$

$$O_3 + HO_2^- \longrightarrow O_3^- + HO_2^{\bullet} \tag{4}$$

$$HO_2^{\bullet} \rightleftharpoons O_2^- + H^+ \tag{5}$$

$$O_3 + O_2^- \longrightarrow O_3^- + O_2 \tag{6}$$

$$O_3^- + H^+ \rightleftharpoons HO_3 \tag{7}$$

$$HO_3 \longrightarrow HO^{\bullet} + O_2 \tag{8}$$

$$HO^• + H_2O_2 \longrightarrow H_2O + HO_2^• \tag{9}$$

$$HO^• + O_3 \longrightarrow O_2 + HO_2^• \tag{10}$$

$$2\,HO^• \longrightarrow H_2O_2 \tag{11}$$

$$2\,HO_2^• \longrightarrow H_2O_2 + O_2 \tag{12}$$

$$H_2O + HO_2^• + O_2^- \longrightarrow H_2O_2 + O_2 + OH^- \tag{13}$$

In AOP's using ultraviolet light, the UV serves as an activator of H_2O_2 but also influences organic compounds as well.

The H_2O_2/UV combination involves one step initiation and the photolytic dissociation of H_2O_2 into two hydroxyl radicals (see Equation 2). It is the most efficient AOP radical producing mechanism with respect to H_2O_2 efficiency. However, the utilization of UV light is relatively inefficient due to the low absorption coefficient of H_2O_2. Even so, UV activated AOP's are viable water treatment processes. The potential for photolytic activation of organic compounds can result in photo-decomposition of the pollutant as well as activating the organic molecule for oxidation by H_2O_2 or hydroxyl radical.

AOP's involving H_2O_2 and ozone combinations present an alternative to H_2O_2/UV systems. While the chemistry has been known for many years [16] its potential in water treatment has only recently been realized [3]. A summary of the reactions involved in an H_2O_2-ozone AOP is as follows (from Glaze, 1987 [9]):

$$H_2O_2 \rightleftharpoons H^+ + HO_2^-$$

$$O_3 + HO_2^- \longrightarrow O_3^- + HO_2$$

$$O_3^- + H^+ \rightleftharpoons HO_3 \qquad HO_2 \rightleftharpoons H^+ + O_2^-$$

$$HO_3 \longrightarrow HO^• + O_2 \qquad O_2^- + O_3 \longrightarrow O_2 + O_3^-$$

$$O_3^- + H^+ \rightleftharpoons HO_3$$

$$HO_3 \longrightarrow HO^• + O_2$$

$$\boxed{H_2O_2 + 2O_3 \longrightarrow 2\,HO^• + 3O_2}$$

This shows that the combination of one H_2O_2 with two ozone molecules results in the formation of two hydroxyl radicals through a combination of several potential steps. One benefit of H_2O_2/ozone AOP's is the lessened impact of turbidity or color of wastewater compared to UV-based AOP's.

The combination of H_2O_2, ozone, and UV in an AOP offers an exciting combination of the beneficial aspects of H_2O_2/UV and H_2O_2/ozone systems [5]. As has been shown [3] the complex chemistry of AOP can involve ozone reactions, H_2O_2 formation and photolysis, various reactions of perhydroxyl radicals, and formation of superoxide. The importance of various pathways and components of the cycle depend on the organics being oxidized. For example, certain organics result in the production of superoxide, which has been said to adequately feed and propagate the reaction of ozone along [13] (Figure 1). Other organics do not produce superoxide, forming instead H_2O_2.

In relatively simple wastewater streams, mechanisms can be developed such that the use of two component AOP's is sufficient for destructive oxidation of the pollutant. However, in complex wastewater streams, both mechanisms can be occurring simultaneously, with numerous chain inhibiting processes also possible. Under these conditions H_2O_2/ozone/UV AOP's can offer an advantage.

Several other advantages of the ternary system include applicability to both clear and colored wastewater streams and to turbid wastewater streams, and operation over a broader pH range. These are of particular interest with respect to reliability of performance under changing stream conditions. For example, in a groundwater treatment application it is possible that groundwater quality can change from time to time. If turbidity increases, the performance of a H_2O_2/UV AOP will suffer markedly. If ozone is also available, however, a smaller impact on treatment efficiency is observed [17].

Application of H_2O_2 AOP's to Wastewater Treatment

The use of hydrogen peroxide in AOP has been implemented in numerous cases commercially, and its utility continues to grow. Because of developing capabilities of H_2O_2 in AOP applications a wider variety of pollutants can be targeted for AOP systems. Additionally, as treatment goals become tighter, the use of combined treatment schemes becomes more attractive. Instead of being the only treatment system, AOP's are effective as pre-treatment or polishing steps. Examples of the broader pollutant scope and of applicability in combined treatment schemes appear below.

MTBE in Groundwater

The remediation of leaking underground gasoline storage tank sites has created a need for treatment technology not only for the characteristic benzene/toluene/xylene (BTX) compounds but also for oxygenated additives such as methyl tert-butyl ether (MTBE). These oxygenates are not easily treated with activated carbon. MTBE, in particular, contains functionalities that are sometimes difficult to treat with AOP - alkane groups and the ether linkage. Two examples of water containing MTBE and BTX are presented.

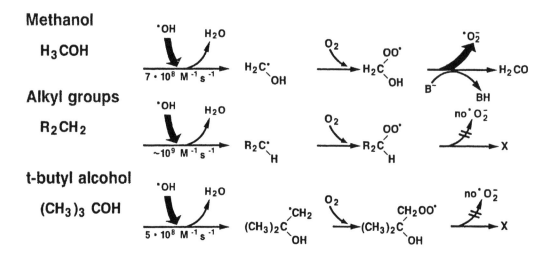

FIGURE 1. Examples of reactions that produce superoxides (methanol) and do not produce superoxides (alkyl groups and t-butyl alcohol), from Stachelin and Hoigne, 1985 [14].

A contaminated groundwater sample containing the levels of BTX and MTBE presented in Table III was investigated in the laboratory. Results of AOP treatment with H_2O_2/ozone/UV are also shown. The aromatic constituents, which are easily oxidized in AOP systems, are more quickly removed than MTBE. However, significant MTBE removal was also observed in most cases, at a rate nearly as fast as the aromatics.

A similar organic component mixture, described in the water analysis in Table IV, in tank water at a gasoline storage terminal was investigated in a pilot plant. The treatment objectives of AOP at this site were the removal of aromatic constituents benzene, toluene, ethylbenzene, and xylenes. However, a particularly high level of MTBE was also present in this water.

AOP oxidation with H_2O_2/ozone/UV showed that the BTEX components were rapidly destroyed to below limits of detection of 5 ppb. As with the groundwater sample, MTBE was also found to undergo considerable oxidation in this system. At the point of BTEX removal to non-detect levels (<5 ppb), 88% of the MTBE was also removed. Continued oxidation demonstrated that non-detectable levels of MTBE could also be achieved.

Based on the pilot results (Table IV), direct operating and maintenance cost for a full scale AOP system to treat the tank storage water are estimated at three cents per gallon. Current disposal methods for tank bottom water (sent to contract waste treatment) resulted in costs in twenty cents per gallon.

The significance destruction effects on all the synthetic organic compounds found in these waters illustrates the potential of H_2O_2 based AOP's in achieving increasingly low levels of organic contaminants in discharged water.

COD Oxidation in Biotreatment Effluent

The use of AOP in combination with other water treatment processes can result in reduced overall treatment costs than when a process is implemented as the sole treatment system. Use of AOP as a polish step in BOD/COD removal exemplifies the potential cost savings that can be gained from H_2O_2/ozone/UV AOP.

A chemical manufacturer currently treating wastewater in a biological treatment step was being forced, under new permit standards, to further reduce BOD and COD in the effluent. A potential post-treatment step was AOP oxidation. Various AOP combinations were evaluated. Laboratory results are presented in Table V for two BOD/COD levels. H_2O_2/ozone/UV reduced COD by more than 70% in 20 minutes reaction time.

Based on the laboratory and pilot work, cost analyses for treating a 650 gpm wastewater steam revealed that the H_2O_2/ozone/UV combination was the most cost-effective AOP (see Table VI). This AOP reduced COD by more than 80% in 20 minutes. Ozone/UV removed only 50% of COD in 30 minutes. With H_2O_2/ozone high removal efficiencies could be achieved but this required a 50% longer reaction time than H_2O_2/ozone/UV. As a result both capital and oxidant costs for the combination of all three AOP components are less than the cost of the AOP alternatives.

In another COD example, wastewater from a fragrance manufacture required reduction of COD to 250 mg/L from 630 mg/L. The results of a pilot study are given in Table VII. H_2O_2/UV achieved only a 35% reduction and ozone/UV a 40% COD reduction in one hour reaction time. H_2O_2/ozone/UV removed 90% of the COD at similar oxidant levels and time. Treatment objectives were met within 30 minutes at less than 70% of the oxidant level.

Conclusions

Advanced Oxidation Processes based upon hydrogen peroxide activation with ozone and/or ultraviolet light are technically and commercially viable water treatment technologies for stand-alone treatment or effluent polishing. The demonstrated capabilities of AOP's allow wide applicability as a cost-effective treatment system. When combined with other treatment processes such as biological treatment, the scope and of AOP can be greatly

TABLE III. Case Study: H₂0₂-Ozone-UV Treatment of BTEX in Groundwater (Laboratory Data)

	O_3 mg/L	H_2O_2 mg/L	MTBE ppb	Benzene ppb	Toluene ppb	Xylenes ppb	Ethyl-Benzene
Raw water	0	0	8670	10833	7000	4830	806
15 minutes	109	217	7	N/D	N/D	N/D	N/D
30 minutes	218	217	N/D	N/D	N/D	N/D	N/D

TABLE IV. Case Study: BTX Oxidation - Tank Bottom Water (Based on Pilot Plant Data Treating 3 gpm with a 120 Minute Retention Time)

Run Number	O_3 mg/L	H_2O_2 mg/L	MTBE ppb	Benzene ppb	Toluene ppb	Xylenes ppb
Pretreated	0	0	180,000	12,000	23,000	10,000
Effluent	700	700	22,000	<5	<5	<5

TABLE V. Case Study: BOD/COD Oxidation with AOP (Based on Pilot Plant Data at 30 gpm)

Run #	Time	H_2O_2 mg/l	O_3 mg/l	COD mg/l	BOD mg/l
1	0	0	0	150	27.0
	20	60.0	187	72	<6.0
2	0	0	0	142	26.0
	20	60.0	187	60.0	<6.0
3	0	0	0	142	77.0
	20	79.0	268	66.0	15.0

TABLE VI. Comparison of AOP Costs for a 650 gpm Operating System

	Retention Time	Oxidant mg/l	Avg. % COD Reduction	$ Capital	$/1000 gals O & M
$H_2O_2/O_3/UV$	20	265	83	1,625,000	$2.15
H_2O_2/O_3	30	360	69	2,195,000	$3.39
O_3/UV	30	360	50	2,415,000	$3.88

a: Current biological treatment plant effluent ranged from 26-77 mg/L BOD while the treatment objective is <20 mg/L BOD.

TABLE VII. Case Study: Frangrance Manufacturer Wastewater - COD Reduction with H_2O_2-Ozone-UV

Run #	Time	H_2O_2 mg/l	O_3 mg/l	COD mg/l
1	0	0	0	630
	30	840	0	500
	60	1080	0	410
2	0	0	0	630
	30	0	660	490
	60	0	1320	390
3	0	0	0	630
	30	480	420	140
	60	480	840	63

a: Activated sludge plant effluent varied from 630-800 mg/L COD while the treatment objective was <275 mg/L COD.

expanded. The resulting combined treatment system can be the most cost-effective option when compared to any of the components used as the sole treatment system.

The utilization of multi-component AOP's can result in significant performance and cost savings compared to H_2O_2/UV of H_2O_2/ozone AOP's, including reduced capital costs, as well as operating and maintenance costs. In addition, these combinations offer added flexibility to the "design" the use of oxidants to meet a given situation, i.e., H_2O_2-ozone, H_2O_2-UV and ozone UV.

References

[1] W.H. Glaze, J-W. Kang, and D.H. Chapin, *Ozone Science and Engineering, 9*:335, 1987.

[2] J. Hoigne, and H. Bader, *Water Research, 10*:377, 1976.

[3] G.R. Peyton, W.H. Glaze, *Environmental Sci. Technol.*, 22:761, 1988.

[4] P.C. Singer, *Journal of American Water Works Association*, page 78, 1990.

[5] US Patents 4,792,407 and 4,849,114 to Ultrox International.

[6] J.T. Barich, EPA Forum on Innovative Hazardous Waste Treatment Technologies, May 15-17, 1990.

[7] N. Lewis et al., *Journal of Air and Waste Management Association, 40*:540, 1990.

[8] D.G. Hegar et al., AWWA Annual Conference and Exposition, June, 1988.

[9] W.H. Glaze, and J-W. Kang, *Ind. Engr. Chem. Res.*, 28:1573, 1989.

[10] G. Ayling and H. Castrantas, *Chemical Engineering, 88*(24):79, 1981.

[11] Farhataziz and A.B. Ross, Research Report PB263198, National Bureau of Standards, Washington, DC, 1977.

[12] E. Keating et al., *Industrial Water Engineering, 15*(7):22, 1978.

[13] G.R. Peyton et al., Research Report 206, University of Illinois at Urbana-Champaign, January, 1987.

[14] J. Staehelin, J. Hoigne, *Environ. Sci. Technol., 19*:1206, 1985.

[15] W.H. Glaze, and J-W. Kang, *Environ. Sci. Technol., 22*:761, 1988.

[16] H. Taube, *Trans. Faraday Soc., 53*:656, 1957.

[17] Heek, Smith, and Perry, ACS Symposium on Emerging Technologies in Hazardous Waste Treatment, June, 1990.

RICHARD J. WATTS
MATTHEW D. UDELL
SOLOMON W. LEUNG

Treatment of Contaminated Soils Using Catalyzed Hydrogen Peroxide

ABSTRACT

The on site and in situ generation of strong oxidants is a potential mechanism for treating refractory organic contaminants in soils. One method of introducing oxidants into soils is the catalyzed decomposition of hydrogen peroxide by iron (II) to form hydroxyl radical, which is commonly known as Fenton's reagent. Hydroxyl radical has over twice the oxidation potential of chlorine and is 25% stronger than ozone. It reacts with organic compounds in aqueous solutions with rate constants of 10^7 to 10^{10} M^{-1} sec^{-1}. Hydroxyl radical is therefore a strong, nonspecific oxidant capable of widespread destruction of organic compounds.

Pentachlorophenol (PCP) was used as a model contaminant in the Fenton's reagent treatment of soils of varying complexity. Mineralization of PCP in a simple two-phase system (silica sand-Fenton's reagent) was demonstrated by the removal of the parent compound (PCP) and the total organic carbon associated with the PCP. In addition, stoichiometric quantities of chloride were recovered at the end of the experiment, which also supports mineralization. Fenton's reagent, when used to treat PCP in natural soils and silica sand, was most effective at pH 2-to-3. In soils of varied organic carbon content, no iron ammendment provided the most efficient reaction (i.e., the greatest ratio of the contaminant degradation rate to the peroxide consumption rate). The efficient reactions occurring with no iron addition in natural soils may have been due to the dissolution of iron minerals which promote catalyzed peroxide decomposition or Fenton-like heterogeneous catalysis occurring on the surfaces of iron minerals.

INTRODUCTION

The uncontrolled disposal of hazardous wastes prior to promulgation of the Resource Conservation and Recovery Act has resulted in the contamination of thousands of sites in the United States [1]. In addition, over 18,000 accidental hazardous materials spills occur each year in the U.S. transportation industry [2].

A large number of these sites which have resulted from uncontrolled and accidental hazardous waste releases are characterized by contaminated surface soils. Excavation and

Richard J. Watts and Matthew D. Udell, Department of Civil and Environmental Engineering, Washington State University, Pullman, Washington, U.S.A.
Solomon W. Leung, Association of American Railroads, 50 F Street N.W., Washington, D.C., U.S.A.

landfill disposal, incineration, and bioremediation have been used to clean up contaminated soils; however, not one process works universally for all sites and all contaminants. For example, on site and in situ bioremediation has been effective in treating soils contaminated with hazardous compounds [3]. However, some waste components are toxic to microorganisms and many are slowly biodegraded; half-lives of up to several months for some compounds are not uncommon during bioremediation efforts [4]. Therefore, new processes are needed to clean up the range of contaminants commonly found at hazardous waste sites [5].

The on site introduction of strong oxidants into contaminated soils may provide a means of destroying biorefractory contaminants over time periods less than required for biological treatment. A documented process for generating strong oxidants in aqueous solutions is Fenton's reagent, which is the reaction of hydrogen peroxide and iron (II) to generate hydroxyl radical (OH•) [6]:

$$H_2O_2 + Fe^{2+} ----> OH\bullet + OH^- + Fe^{3+} \tag{1}$$

Hydroxyl radical is second only to fluorine in oxidation potential and is capable of nonspecific oxidations because it reacts with organic compounds with bimolecular rate constants of 10^7 to 10^{10} L/mole·sec [7]. Fenton's reagent also involves numerous competing reactions [6, 8, 9]:

$$H_2O_2 + Fe^{3+} -----> Fe^{2+} + H^+ + HO_2\bullet \tag{2}$$
$$RH + OH\bullet ----->H_2O + R\bullet \tag{3}$$
$$Fe^{2+} + OH\bullet -----> Fe^{3+} + OH^- \tag{4}$$
$$R\bullet + Fe^{3+} -----> Fe^{2+} + products \tag{5}$$
$$R\bullet + OH\bullet -----> ROH \tag{6}$$
$$R\bullet + H_2O_2 -----> ROH + OH\bullet \tag{7}$$
$$HO_2\bullet + Fe^{3+} -----> O_2 + Fe^{2+} + H^+ \tag{8}$$
$$OH\bullet + H_2O_2 -----> HO_2\bullet + H_2O \tag{9}$$

where RH represents an oxidizable substrate, R• is an alkyl radical and $HO_2\bullet$ is superoxide radical. The rate constant for Equation 1 is 76 L/mole·sec. Rate constants for Equations 3, 5, 6, and 7 are substrate specific [8]. The rate constant for Equation 4 is 3×10^8 L/mole·sec [6]; the rate constant for Equation 8 is pH dependent and ranges from 2×10^4 to 1×10^6 L/mole·sec [9, 10].

The fundamental aspects of Fenton's chemistry have been well documented. For example, Walling and Johnson [9] investigated the conditions that favor hydrogen abstraction vs. hydroxyl radical addition to substituted benzenes. Ingles 8] provided similar information with low molecular weight aliphatic alcohols.

Fenton's reagent has recently been applied to the treatment of organic contaminants. Barbeni et al. [12] investigated the Fenton's oxidation of di- and tri-chlorophenols in aqueous solution. A mass balance in their system involving the measurement of residual chlorophenol, chloride, and total organic carbon showed that the chlorophenols were mineralized by the treatment. Fenton's reagent also successfully oxidized a formaldehyde waste stream under bench-scale conditions [13]. The process has also been used to treat wastewaters containing sodium dodecylbenzenesulfonate [14], p-toluenesulfonic acid and p-nitrophenol [15], and azo dyes [16]. However, few studies have systematically investigated the Fenton's reagent treatment of contaminated soils. As part of a comprehensive effort to study the Fenton's reagent treatment of soils, the purpose of our research has been to determine the optimum conditions for Fenton's soil treatment. Most Fenton's reagent research has used H_2O_2 concentrations from 0.05% to 1.0% [9]. However, this investigation used H_2O_2 concentrations in excess of what would be required to stoichiometrically oxidize the substrate in order to promote mineralization.

METHODS AND MATERIALS

MODEL CONTAMINANT

Pentachlorophenol, a widely-used wood preservative, was used as a model contaminant. Pentachlorophenol is moderately biorefractory. Mabey et al. [17] reported a biodegradation rate constant of 3×10^{-12} L/cell·hour, log K_{ow} of 5.01, and vapor pressure of 1.1×10^{-4} mm Hg at 20°C.

SOILS

Commercially-available silica sand (40-100 mesh) and one natural soil were used. The natural soil was a grayish-brown, gravelly-loamy coarse sand, mixed, mesic, torriothentic haploxeroll. The soil, which was sampled from an alluvial fan in the Carson Valley, Nevada, is one of low development; therefore, successive horizons provided a gradient of soil organic carbon contents, but were relatively uniform in particle size distribution, mineralogy, cation exchange capacity, and pH. Particle size analysis was determined by the pipette method [18]. Organic carbon was determined by combustion at 900°C with evolved CO_2 trapped in KOH and measured by back titration of unreacted KOH [19]. Amorphous and crystalline iron and manganese oxyhydroxides were determined by citrate-bicarbonate-dithionite extraction [20]. Cation exchange capacity was established by saturation with sodium acetate at pH 8.2 [21]. The soil characteristics are shown in Table 1.

PROCEDURES

The soils were spiked with 250 mg/Kg pentachlorophenol by adding a PCP-acetone solution to the soil; the acetone was allowed to evaporate. Experiments were conducted in batch, completely mixed systems. To 2.5 g of PCP-contaminated soil in a 40 mL borosilicate glass vial, 12.5 mL of 7% H_2O_2 were added followed by 1 mL of a $FeSO_4$ solution or deionized water to provide a final ammendment of 0, 240 mg/L (0.024%), or 400 mg/L (0.040%) iron in the aqueous phase.

Separate experimental units conducted in duplicate vials were used to monitor treatment effectiveness at a minimum of 5 time periods over 24 hours. At the time designated, an aliquot of the soil-water was collected for hydrogen peroxide analysis. The reaction was then stopped by the addition of eight drops of concentrated sulfuric acid to the vials. The ability of concentrated H_2SO_4 to quench PCP oxidation was documented by monitoring pentachlorophenol degradation in acidified samples over 8 hours. No PCP degradation was observed. The soil slurries were shake-extracted with 3 mL of ethyl acetate for 30 minutes and then centrifuged for 15 minutes at 1,200 rpm.

Soluble iron was measured in parallel samples by filtering the soil slurry through a GFC filter. Total organic carbon and chloride were also measured on experiments conducted in parallel. Total organic carbon was determined by conducting the reactions in sealed TOC ampules. At selected times, evolved CO_2 was measured on an OI 700 TOC analyzer.

TABLE I. SOIL CHARACTERISTICS

Organic carbon content (%)	0.2	0.5	1.0	1.7
% sand	85.3	86.5	86.3	86.1
% silt	12.3	11.0	10.9	10.8
% clay	2.4	2.5	2.8	3.1
Crystalline Fe oxides (%)	0.44	0.44	0.43	0.43
Crystalline Mn oxides (%)	0.01	0.01	0.01	0.01
Amorphous Fe oxides (%)	0..34	0.44	0.42	0.40
Amorphous Mn oxides (%)	0.01	0.01	0.01	0.01
Cation exchange capacity cmol/kg	4.04	4.28	4.59	4.90
pH	6.6	6.4	6.5	6.6

ANALYSES

Hydrogen peroxide concentrations were determined by iodometric titration with sodium thiosulfate. Residual pentachlorophenol was followed by gas chromatography after ethyl acetate extraction from the soil. A Hewlett-Packard 5890A gas chromatograph with a flame ionization detector and a 0.53 mm (ID) x 15m Supelco SPB-5 capillary column was used. Chromatographic conditions were: initial oven temperature 100°C; final oven temperature 240°C; program rate 30°C/min; injector temperature 200°C; detector temperature 240°C; and nitrogen carrier gas flow rate 20 mL/min. Soluble iron was analyzed using flame methodology on a Perkin Elmer 3030B atomic absorption spectrophotometer. Chloride was determined using a Fisher chloride electrode paired with a double junction reference electrode.

RESULTS AND DISCUSSION

TREATMENT KINETICS AND STOICHIOMETRY

Hydrogen peroxide consumption and pentachlorophenol degradation during a typical treatment experiment (250 mg/Kg pentachlorophenol, 7% hydrogen peroxide, and pH 3) as a function of time are shown in Figure 1. The data of Figure 1 show that pentachlorophenol and hydrogen peroxide first decomposed rapidly, with a slower decomposition rate after 3 hours of reaction. The expected rate in a Fenton's system is zero order because hydroxyl radical generation should approach steady state (Equation 1) with pentachlorophenol present in high concentrations; however, all decomposition rates, regardless of the hydrogen peroxide and iron ammendments, were not linear. The soluble iron concentrations as a function of time are also shown in Figure 1. Only 23% of iron added to the system could be recovered in solution at time zero. In addition, the soluble iron concentration decreased over the first three hours of treatment and the concentration remained relatively constant thereafter. A possible mechanism for iron precipitation may be:

$$Fe^{2+} + 1/2\,O_2 + 2OH^- \longrightarrow \gamma\text{-}FeOOH + H_2O \qquad\qquad (10)$$

which has a reaction half-life of 25 minutes at neutral pH [22]. The concentration of OH^- in the Fenton's system at pH 3 is four orders of magnitude less than in a system at neutral pH; a half-life adjusted for the difference in OH^- concentration may explain the 3 hour time requirement to reach the quasi-equilibrium soluble iron concentration of 9 mg/L.

The importance of soluble iron in catalyzing Fenton's reactions has been well documented [6]. Because the concentration of pentachlorophenol, peroxide, and soluble iron changed over time in the silica sand systems, rate quantitation was difficult. A turnover number commonly used in catalysis could not be used because the soluble catalyst concentration decreased over the course of the reaction. Therefore, pentachlorophenol and hydrogen peroxide concentrations as a function of time were fit to zero, first, and second-order models [23]. The first-order model provided the best fit of the data with $R^2 > 0.90$ for plots of the natural logarithm of concentration as a function of time. First-order fit of experimental data is a common practice for quantifying complex environmental processes, such as biochemical oxygen demand [24]. The reactions occurring in the Fenton's system are complex; however, the empirical fit of experimental data to the first-order model provided the most accurate means of comparing different treatment conditions.

The concentrations of pentachlorophenol, total organic carbon, and chloride over the course of the reaction followed in Figure 1 are shown in Figure 2. These data show that greater than 99.9% of the original pentachlorophenol was degraded in 24 hours and that the removal of total organic carbon closely followed pentachlorophenol degradation. These results are similar to those obtained by Barbeni et al. [12] who demonstrated the mineralization of di- and tri-chlorophenols in aqueous systems. The phenomenon that organic carbon was removed rapidly after pentachlorophenol degradation suggests that hydroxyl radical attack on the products is more rapid than on the parent compound. High rates of product degradation may be explained by the lower oxidation state of the ring as it is hydroxylated and the increased water solubility of the products. Substrate water solubility has been implicated in Fenton's reagent treatment reactivity [25]. The high oxidation state of halogenated organics also significantly affects reactivity with oxidizing species [26].

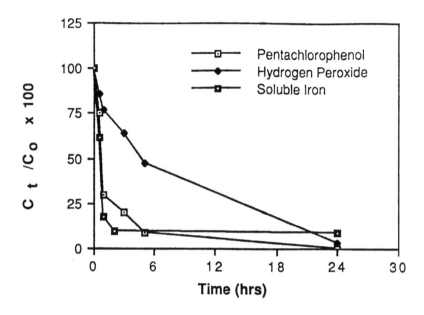

Figure 1. Concentrations of pentachlorophenol, hydrogen peroxide, and soluble iron during Fenton's treatment of 250 mg/Kg PCP in silica sand with 0.040% iron additon.

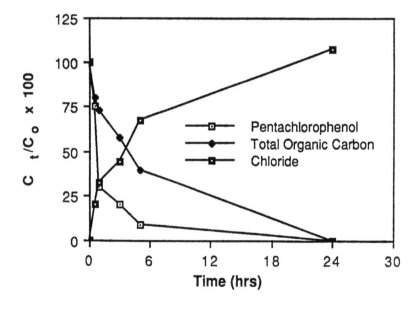

Figure 2. Concentrations of pentachlorophenol, total organic carbon, and chloride during Fenton's treatment of 250 mg/Kg PCP in silica sand.

EFFECT OF TREATMENT pH

First-order rate constants for PCP degradation and H_2O_2 decomposition as a function of pH in silica sand are shown in Figure 3 for experiments conducted with 0.040% iron addition. Treatment of pentachlorophenol in silica sand with no iron addition showed no degradation over 24 hours. With iron addition, PCP degradation rate constants increased significantly at low pH. Pentachlorophenol degradation was evident above pH 6 but at lower rates. The higher PCP degradation rate constants at low pH are probably related to the requirement of soluble iron in the system (6). In addition, the pH regime may enhance the cycling of iron (II) as described by Equations. 2, 5, and 8. While hydroxyl radical generation is enhanced at low pH, oxygen evolution is the predominant route of H_2O_2 decomposition at neutral pH [27]:

$$H_2O_2 \text{-------}> 1/2 \, O_2 + H_2O \qquad (11)$$

The optimum treatment conditions are characterized by a high rate of contaminant degradation with a minimal rate of H_2O_2 consumption. The data of Figure 3 show that the highest PCP degradation rate constants and the lowest peroxide decomposition rate constants were at pH 2 and 3. Similar results were obtained for reactions conducted in natural soils. In some cases PCP degradation rate constants at pH 4, 5, and 6 were higher than rate constants at pH 2 and 3. However, the corresponding rate constants for H_2O_2 decomposition increased above pH 3, which resulted in lower treatment efficiencies. Therefore, the optimum pH regime for all of the soil systems investigated was in the range of pH 2-to-3.

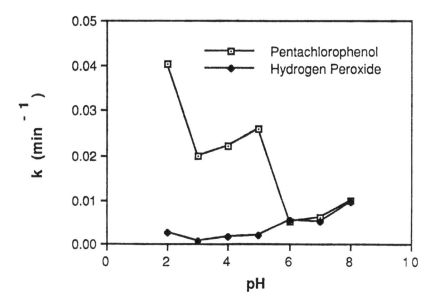

Figure 3. Effect of pH on the first-order rate constants for PCP degradation and H_2O_2 consumption.

First order rate constants for PCP degradation as a function of soil organic carbon are shown in Figure 4 for experiments conducted with 0, 0.024%, and 0.040% iron ammendments. Treatment of soils with 0.2% and 0.5% organic carbon and iron addition resulted in undetectable pentachlorophenol concentrations after 3 days. A factorial design analysis of variance (ANOVA) was used to evaluate equivalence of rate constant population means for the soils with the four soil organic carbon contents when iron was added. Pentachlorophenol degradation rate constants were significantly greater ($\alpha \leq 0.05$) in the soils with organic carbon < 1.0%. Degradation rate constants for pentachlorophenol degradation were not significantly different ($\alpha \leq 0.05$) when the soil organic carbon contents were $\geq 1.0\%$. These data suggest that soil organic carbon competed with PCP for hydroxyl radical generated by catalyzed peroxide decomposition.

Figure 5 shows the H_2O_2 decomposition rate constants as a function of soil organic carbon content for the systems with 0, 0.024% and 0.040% iron additions. Statistical analysis (ANOVA) showed that hydrogen peroxide decomposition rates were greater in systems with the addition 0.040% iron than in systems with 0.024% iron. Peroxide decomposition rates were also greater in systems with iron addition than systems without iron addition ($\alpha \leq 0.05$). In addition, no significant difference ($\alpha \leq 0.05$) was found between H_2O_2 decomposition rates as a function of soil organic carbon when no iron was added.

The optimum treatment efficiency in a Fenton's system is characterized by maximum contaminant degradation and minimum peroxide consumption. A ratio of contaminant degradation rate constants (k_{PCP}) to H_2O_2 decomposition rate constants ($k_{H_2O_2}$) is therefore an empirical measure of the relative efficiency of Fenton's reagent treatment. The ratios were

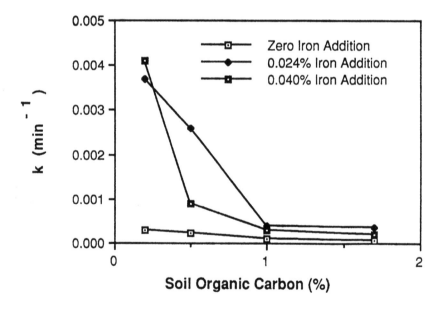

Figure 4. First-order rate constants for PCP degradation as a function of soil organic carbon.

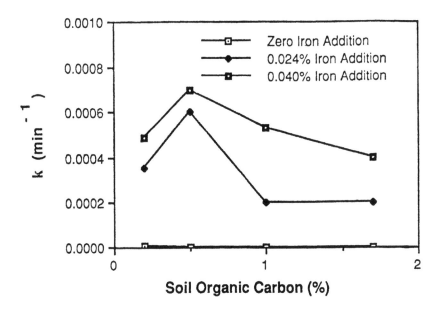

Figure 5. First-order rate constants for hydrogen peroxide consumption as a function of soil organic carbon

greater in the systems that did not receive the iron ammendment (Table II). In addition, the $k_{PCP}/k_{H_2O_2}$ ratios were higher in the soils with lower organic carbon contents for the treatment of pentachlorophenol. The data of Table II show that Fenton's treatment efficiencies were not only sensitive to the soil organic carbon content, but also to the iron ammendment. Therefore, the most efficient and economical Fenton's reagent treatment of PCP occurred without the addition of iron at low soil organic carbon contents.

The results of this research show that a range of biorefractory contaminants can be treated in soils at pH 3 in the presence of H_2O_2 without iron addition. The highest degradation rates for PCP occurred with the addition of iron; however, the hydrogen peroxide decomposition rates were also higher, which resulted in low treatment efficiencies. The process was most efficient when no iron was added to the system for all soil organic carbon contents evaluated. The PCP degradation efficiencies in natural soils (i.e., optimum treatment conditions requiring no iron amendment) suggest that crystalline and amorphous iron oxides may be promoting catalyzed peroxide decomposition.

Table II. ESTIMATION OF FENTON'S TREATMENT EFFICIENCY

Iron Addition	Organic Carbon			
	0.2%	0.5%	1.0%	1.7%
	$k_{PCP}/k_{H_2O_2}$			
0.040%	6.2	2.2	0.30	0.47
0.024%	13	1.5	0.59	1.0
0.00%	40	13	10	5.4

To determine the availability of iron minerals to catalyze Fenton-like oxidations, 10% mixtures of goethite (α-FeOOH), hematite (Fe_2O_3), and magnetite (Fe_3O_4) in silica sand were spiked with 250 mg/Kg PCP. The goethite, hematite, and magnetite systems contained 6.3%, 7.0%, and 7.5% iron, respectively. Fenton's reactions were conducted in these systems at pH 3 with 7% H_2O_2; the mineral was the only source of iron. Figure 6 shows that the presence of goethite, hematite, and magnetite in the silica sand systems resulted in loss of PCP over 24 hours. Control experiments with no mineral addition showed no PCP degradation. To confirm PCP degradation in the mineral systems, total organic carbon and chloride were monitored in conjunction with PCP residual over 24 hours of treatment. The results of magnetite catalyzed oxidation of PCP are shown in Figure 7. These data show that PCP was mineralized with stoichiometric recovery of chloride and removal of organic carbon to below detectable levels (< 1 μg). Pentachlorophenol was oxidized to CO_2, H_2O, and HCl as in the Fenton's systems containing soluble iron (Figure 2). Mineralization of PCP in goethite-silica sand and hematite-silica sand systems was also demonstrated. The relationship of the mass of hematite in a silica sand system to the degree of Fenton-like PCP degradation is illustrated in Figure 8. These data show that PCP degradation was directly proportional to the weight percentage of hematite in the system.

The ability of minerals to promote Fenton-like oxidations may occur by at least two mechanisms: 1). mineral dissolution with release of soluble iron which then catalyzes peroxide decomposition, and 2). heterogeneous catalysis on mineral surfaces. Our research is currently investigating these mechanisms in mineral systems and natural soils with emphasis on optimum treatment efficiencies.

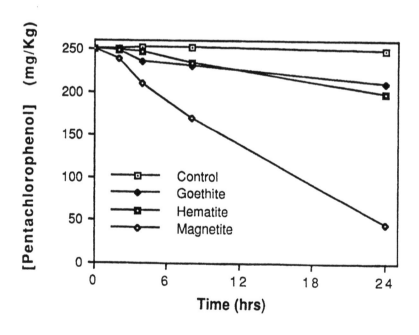

Figure 6. Pentachlorophenol degradation with minerals as the sole iron source.

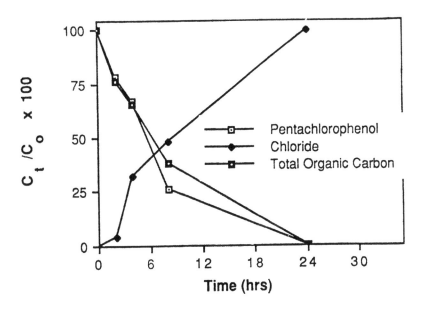

Figure 7. Concentrations of pentachlorophenol, total organic carbon, and chloride during Fenton's treatment of 250 mg/Kg PCP in silica sand with 16.7% magnetite as the sole iron source.

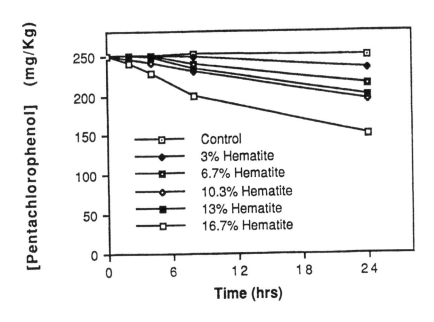

Figure 8. Degradation of pentachlorophenol with hematite as the sole iron source in silica sand.

Because of the complexity of Fenton's treatment systems, bench-scale treatability studies will probably be necessary to assess the efficacy of Fenton's reagent treatment on a site by site basis using a matrix of pH regimes and iron ammendments [28]. Such bench-scale treatability studies are commonplace in assessing the potential for bioremediation of surface soils and groundwater [4]. Further Fenton's research will provide predictive treatability models based on chemical structure-reactivity, catalysis by iron minerals, and the effect of soil organic carbon.

CONCLUSIONS

1. The optimum regime for Fenton's treatment of contaminated soils was pH 2-to-3.

2. In successive horizons of the natural soil investigated, the effects of Fenton's treatment was sensitive not only to organic carbon, but also iron addition.

3. The optimum treatment efficiency was demonstrated with no iron addition.

4. The ability of iron minerals to catalyze Fenton-like reactions provided efficient treatment with maximum PCP degradation and minimal hydrogen peroxide consumption.

AKNOWLEDGEMENTS

Funding for this project was provided by the U.S. Environmental Protection Agency through Assistance Agreement No. R-814425-01-0. Additional support was provided by Interox America and the National Science Foundation through Engineering Research Equipment Grant No. CES-8704878. We thank Dr. Robert Monsen for useful discussions of peroxide chemistry. We also thank Bryan Tyre and Paul Rauch for technical assistance.

REFERENCES

1. Exner, J.H. 1987. Perspective on Hazardous Wastes Problems Related to Dioxins. In: *Solving Hazardous Waste Problems. Learning from Dioxins*. J.H. Exner, Ed. ACS Symposium Series 338. Washington, D.C.

2. Ryckman, D.W., and Ryckman, M.D. 1980. Organizing to Cope with Hazardous Material Spills. *Jour. Amer. Water Works Assn.* 72: 196-200.

3. Thomas, J.M., and Ward, C.H. 1989. In Situ Biorestoration of Organic Contaminants in the Subsurface. *Environ. Sci. Technol.* 23: 760-766.

4. Lyman, W.J., Reehl, W.F., and Rosenblatt, D.H. 1982. *Handbook of Chemical Property Estimation Methods*. McGraw-Hill. New York.

5. U.S. Environmental Protection Agency. 1985. *Remedial Actions at Waste Disposal Sites.*. EPA/625/6-85/006. Washington, D.C.

6. Walling, C. 1975. Fenton's Reagent Revisited. *Acc. Chem. Res.*. 8: 125-131.

7. Dorfman, L.M., and Adams, G.E. 1973. *Reactivity of the Hydroxyl Radical*. National Bureau of Standards Report No. NSRDS-NBS-46.

8. Ingles, D.L. 1972. Studies of Oxidations by Fenton's Reagent Using Redox Titration. I. Oxidation of Some Organic Compounds. *Aust. J. Chem..* 25: 87-95.

9. Walling, C., and Johnson, R.A. 1975. Fenton's Reagent. V. Hydroxylation and Side-chain Cleavage of Aromatics. *Jour. Am. Chem. Soc.* 97: 363-367.

10. Yang, T.C., and Neely, W.C. 1984. Relative Stoichiometry of Oxidation of Ferrous Ion by Ozone in Aqueous Solution. *Anal. Chem.* 58: 1551-1555.

11. Staehelin, J. 1984. Ozone Decomposition in Water Studied by Pulse Radiolysis. 2. OH and HO_4 as Chain Intermediates. *Jour. Phys. Chem.* 88: 5999-6004.

12. Barbeni, M., Minero, C., Pelizzetti, E., Borgarello, E., and Serpone, N. 1987. Chemical Degradation of Chlorophenols with Fenton's Reagent. *Chemosphere.* 16: 2225-2237.

13. Murphy, P., Murphy, W.J., Boegli, M., Price, K., and Moody, C.D. 1989. A Fenton-like Reaction to Neutralize Formaldehyde Waste Solutions. *Environ. Sci. Technol.* 23: 166-169.

14. Sato, S., Kobayashi, T., and Sumi, Y. 1975. Removal of Sodium Dodecylbenzene Sulfonate with Fenton's Reagent. *Yukagaku.* 24: 863-868.

15. Feuerstein, W., Gilbert, E., and Eberle, S.H. 1981. Model Experiments for the Oxidation of Aromatic Compounds by Hydrogen Peroxide in Wastewater Treatment. *Vom Wasser.* 56: 35-54.

16. Kitao, T., Kiso, Y., and Yahashi, R. 1982. Studies on the Mechanism of Decolorization with Fenton's Reagent. *Mizii Shori Gijutsu,.* 23: 1019-1026.

17. Mabey, W.R. 1982. *Aquatic Fate Process Data for Organic Priority Pollutants.* EPA 440/4-81-014.

18. Gee, B.W., and Bauder, J.W. 1986. Particle-size Analysis. In: *Methods of Soil Analysis. Part 1. Physical and Mineralogical Methods.* A. Klute, Ed. pp. 399-404. American Society of Agronomy. Madison, Wisconsin.

19. Nelson, D.W., and Sommers, L.E. 1982. Total Carbon, Organic Carbon, and Organic Matter. In: *Methods of Soil Analysis. Part 2. Chemical and Microbiological Properties,.* A.L. Page, Ed. pp. 539-579. American Society of Agronomy. Madison, Wisconsin.

20. Jackson, M.L., Lim, C.H., and Zelazny, L.W. 1986. Oxides, Hydroxides, and Aluminosilicates. In *Methods of Soil Analysis. Part 1. Physical and Mineralogical Methods.* A. Klute, Ed. pp. 113-124. American Society of Agronomy. Madison, Wisconsin.

21. Soil Conservation Service. 1972. *Soil Survey Laboratory Methods and Procedures for Collecting Soil Samples,.* Soil Survey Investigation: Report 1. U.S. Government Printing Office. Washington, D.C.

22. Sung, W., and Morgan, J.J. 1980. Kinetics and Product of Ferrous Iron Oxygenation in Aqueous Systems. *Environ. Sci. Technol..* 14: 561-568.

23. Fogler, H.S. 1986. *Elements of Chemical Reaction Engineering*. Prentice-Hall. Englewood Cliffs, New Jersey.

24. Tchobanoglous, G., and Schroeder, E.D. 1985. *Water Quality*, Addison-Wesley Publishing. Reading, Massachusetts.

25. Sheldon, R.A., and Kochi, J.K. 1981 *Metal-Catalyzed Reactions of Organic Compounds: Mechanistic Principles and Synthetic Methodology Including Biochemical Processes*. Academic Press, New York.

26. Harris, A.H. 1988. *Methods for the Oxidation of Organic Compounds: Alcohols, Alcohol Derivatives, Alkyl Halides, Nitroalkanes, Alkyl Azides, Carbonyl Compounds, Hydroxyarenes and Aminoarenes*. Academic Press. San Diego, California.

27. Dixon, W.T., and Norman, R.O.C. 1962. Free Radicals Formed During the Oxidation and Reduction of Peroxides. *Nature*. 196: 891-892.

28. Watts, R.J., M.D. Udell, P.A. Rauch, and Leung, S.W. 1990. Treatment of Pentachlorophenol-Contaminated Soils Using Fenton's Reagent. *Haz Waste Haz Mater*. 7:335-345.

S. F. ROBINSON
R. M. MONSEN

Hydrogen Peroxide and Environmental Immediate Response

ABSTRACT

Several case histories are discussed with respect to reaction kinetics, dosing and equipment used with hydrogen peroxide. The focus of the paper is characterization of those conditions which favor the use of hydrogen peroxide for environmental emergency response.

INTRODUCTION

Increased and more stringent environmental regulations coupled with the ever-quickening pace of a technological society are redefining our definitions of an environmental emergency. Common practices of a few years ago are now considered highly undesirable and often illegal. Environmental regulations often require immediate response to temporary situations. These temporary situations can arise from treatment system upsets, seasonal low flow conditions or small accidental spills.

Before initiating an immediate response to an environmental situation the following considerations should be reviewed. A qualitative and quantitative assessment of the chemical parameters responsible for the situation. A review of those technologies which are potential solutions to the problem. A review of all regulatory, safety and financial considerations which apply to the situation.

HYDROGEN PEROXIDE

Hydrogen peroxide has been used for environmental applications for many years.[1,2,3,4] Systems are in place worldwide providing cost-effective treatment of air, water, wastewater and hazardous wastes. Over the past few years, however, the value of hydrogen peroxide for immediate response has been discovered. Figure 1 displays those situations where chemical oxidation (H_2O_2) has the greatest effectiveness. Chemical oxidation with hydrogen peroxide involves optimization of several controlling variables.

S. F. Robinson and R. M. Monsen, Interox America, Houston, Texas, U.S.A.

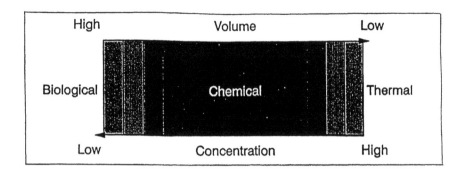

Figure 1. Competitive Technologies

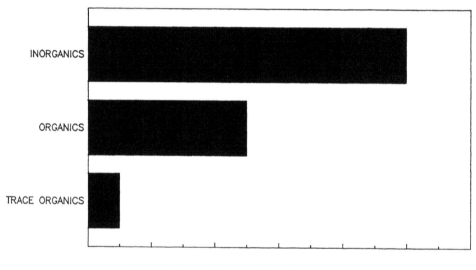

INCREASING RATES OF REACTION —————————>

Figure 2. Relative Oxidation Rates with H_2O_2

These include pH, catalyst, temperature, contact time, application rate and reactivity of compounds. The optimum value for these factors varies with the compound being oxidized. Knowledge of the controlling variables for peroxide oxidation provides a wide degree of flexibility and versatility in its use. Figure 2 relates the relative reactivity of various pollutants with hydrogen peroxide. In general the relative rate of reaction with hydrogen peroxide proceed such that inorganics react faster than organics and trace organics react even slower due to mass transfer limitations.

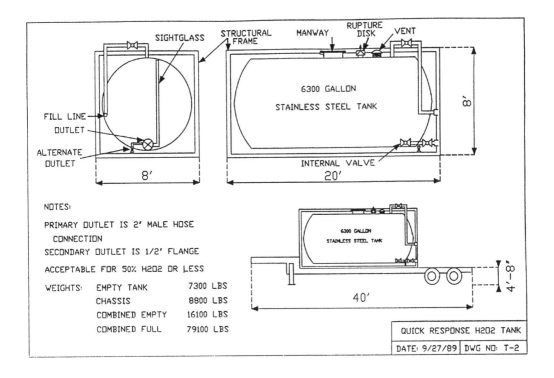

Figure 3. Bottom Unloading Isocontainer

Recent deployment of bottom unloading isocontainers (Figure 3) has had a significant impact on the capability of hydrogen peroxide to be utilized for immediate response. Since these units require no electricity, they can be placed in remote locations to maximize treatment results.

To demonstrate the use of hydrogen peroxide, several case histories will be reviewed with respect to the following:

1. problem definition
2. chemistry and kinetics involved
3. theoretical or lab scale predicted results
4. field set up and dosing
5. actual results attained

TEMPORARY TREATMENT SYSTEM UPSETS

OIL REFINERY CASE HISTORY

A southwestern oil refinery (Figure 4) operates a ten acre wastewater basin which developed a severe odor problem. Field testing revealed the water contained between 5 and 13 mg/L of sulfide, 25 mg/L of thiosulfate and had a pH of 7.5. Chemetrics field test kits employing a methylene blue colorimetric determination of sulfide and an Iodate-Iodide titrimetric determination of thiosulfate were used for analyses. At pH 7.5 most of the sulfide is present as soluble HS⁻, which reacts rapidly with hydrogen peroxide.

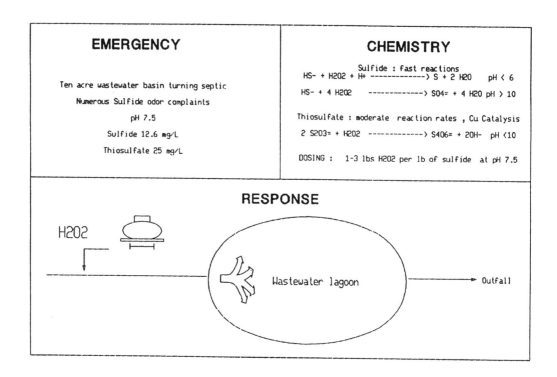

Figure 4. Sulfide Odor Control in A
Refinery Wastewater Lagoon

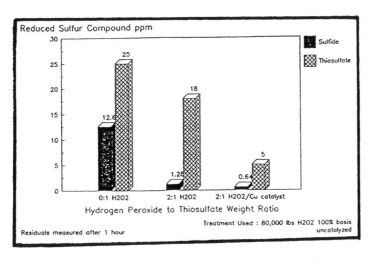

Figure 5. Sulfide Odor Control With H_2O_2
Refinery Wastewater Lagoon

A wastewater sample from the basin was tested under lab conditions which showed that the addition of 2:1 H_2O_2:S_2O_3 w/w removed 90% of the sulfide and 28% of the thiosulfate in one hour (Figure 5). A second sample treated with a 2:1 w/w ratio and using a copper catalyst removed 95% of the sulfide and 80% of the thiosulfate in one hour (Figure 5). Since addition of a copper catalyst was environmentally unacceptable and the retention time of the basin was several days, a 2:1 w/w ratio of uncatalyzed H_2O_2 to thiosulfate was deemed sufficient to treat the lagoon. A bottom unloading ISO (a self contained H_2O_2 delivery and storage vessel), was delivered to the site within two days of the initial call. Figure 5 displays the results of the lagoon treatment with hydrogen peroxide. Due to the seasonal and periodic nature of this odor problem, additional hydrogen peroxide was kept on site as a safeguard for future upsets.

LANDFILL CASE HISTORY

A large industrial landfill operation (Figure 6) in England was cited with a Prohibition Notice under the Public Health Act (Recurring Nuisances) to prevent the production of offensive odors which were emanating from four leachate lagoons [5]. Additionally, the Thames River Authority imposed discharge criteria for the effluent which flowed into the Thames River.

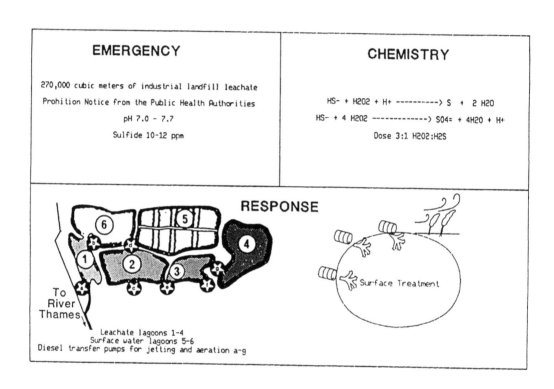

EMERGENCY

270,000 cubic meters of industrial landfill leachate

Prohibition Notice from the Public Health Authorities

pH 7.0 – 7.7

Sulfide 10-12 ppm

CHEMISTRY

$HS^- + H_2O_2 + H^+ \longrightarrow S + 2 H_2O$

$HS^- + 4 H_2O_2 \longrightarrow SO_4^= + 4H_2O + H^+$

Dose 3:1 H_2O_2:H_2S

RESPONSE

Surface Treatment

To River Thames

Leachate lagoons 1-4
Surface water lagoons 5-6
Diesel transfer pumps for jetting and aeration a-g

Figure 6. Sulfide Odor Removal from a Landfill Leachate Lagoon

The most urgent problem for the operation was to eliminate the hydrogen sulfide odor from the leachate lagoons while heavy equipment was being installed to treat and recirculate the leachate. To achieve this objective a well-tried technique of surface treatment with hydrogen peroxide was employed. Fifty-Kg containers of 35% H_2O_2 were placed every 20 meters along the upwind bank of the lagoons. Peroxide was then dripped onto the water at approximately 100 mL/min from each drum. Lagoon 4 was surface treated at 14 day intervals because it was the closest to the site perimeter and was upwind of the source of odor complaints. The complete site treatment strategy was to dose 35% hydrogen peroxide at a 3:1 H_2O_2:H_2S ratio (w/w as 100%) followed by jetting for aeration and mixing. Lagoons 1 and 2 contained 3 mg/L of sulfide while lagoons 3 and 4 contained >10 mg/L of sulfide. After dosing with peroxide all detectable sulfide was removed from the four lagoons. Figure 7 details the results of the leachate treatment for the discharge criterion parameters set forth by the Thames River Authority. Hydrogen peroxide was shown to be an effective treatment of leachate as it destroyed sulfide odors, removed black iron sulfide, maintained aerobic conditions for biological oxidation, and kept discharge ditches and drains free of microbial growths.

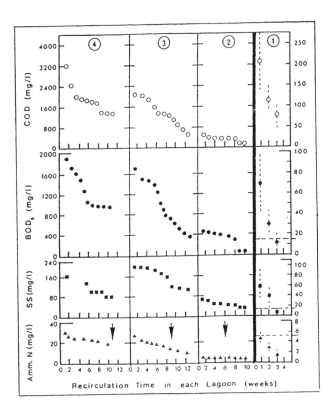

Figure 7. Landfill Leachate Quality Improvement

REFINERY CASE HISTORY

A Gulf Coast refinery (Figure 8) maintained a catalyst waste holding pond on its property. The pond was clay lined and held 20 million gallons of wastewater. The phenol concentration in the waste was 1000 ppm. The pond had developed a weak section in one of its containment dikes. In order to repair the dike, 2 feet of wastewater (8 million gallons) was to be drained from the pond. A Fenton's type treatment system was installed to remove the phenols to a target level of 5 ppm in the polishing pond. Since the wastewater pH was 6.7, no pH adjustment was made. Intermediate dicarboxylic

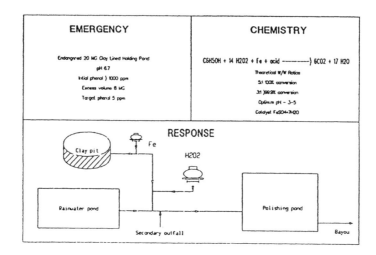

Figure 8. Phenolic Removal from a
Refinery Wastewater Lagoon

acids produced by phenol oxidation with hydrogen peroxide lowered the pH of the wastewater towards the optimum range (pH 3 to 5). A 3:1 H_2O_2:phenol (w/w ratio as 100%) catalyzed with Fe^{+2} was predicted to yield >99.9% reduction of phenol. Figure 9 displays the results of the treatment. Actual results showed that a 3.7:1 H_2O_2:phenol ratio was used to maintain the polishing pond phenol level below 5 ppm. This excess H_2O_2 was dosed initially to assure adequate removal of phenol until the system could be optimized.

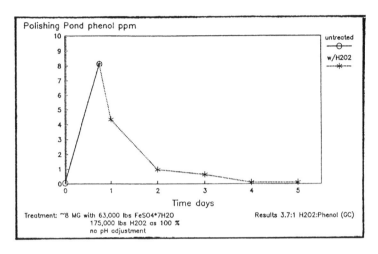

Figure 9. Phenolic Removal with H_2O_2
Endangered Clay Lined Pit/Refinery Waste

PULP AND PAPER CASE HISTORY

A northwestern pulp and paper company diverts process upset wastewater into a ten million gallon holding pond (Figure 10). The plant operates a 40 MGD conventional activated sludge system. The biological treatment system was experiencing settling problems in their secondary clarifiers. The holding pond was being used to divert wastewater in an effort to isolate the cause of the settling problems. As the pond was nearing capacity, wastewater from the pond needed to be pumped into the treatment system. Inhibition studies of the wastewater in the holding pond showed it to be very inhibitory to their biomass. The wastewater in the pond had a pH of 7.2 and a strong organic odor. The COD of the waste was 875 ppm and the TOC was 300 ppm. After reviewing several options for treating the waste it was decided that iron-catalyzed peroxide (Fenton's system) would be used to oxidize the waste in one large batch treatment. Although the exact organic compounds were not specified, the reaction of H_2O_2 and Fe with organics could be represented by the reactions shown in Figure 10.

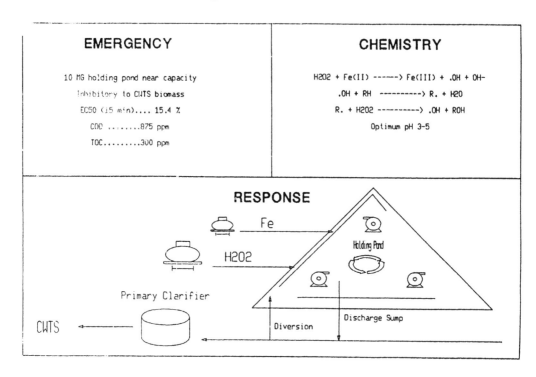

Figure 10. Toxicity Reduction of a Pulp & Paper Mill Wastewater

Since the exact reaction with peroxide was unknown, dosing was based on the TOC at 4-5:1 H_2O_2:TOC w/w ratio. This ratio was a conservative estimate based on the assumption the TOC would react similarly to phenol. The optimum pH for this reaction is 3-5. Sulfuric acid (49,575 lbs 93%) was used to lower the pH to 4.5 to 5 before the iron was added (50,000 lb $FeSO_4 \cdot 7H_2O$). The iron concentration of the pond was increased to 125 ppm before hydrogen peroxide (110,000 lbs as 100%) was added. Mixing in the triangular pond was achieved using three large recirculating pumps at strategic points around the pond. Chemical addition began on a Tuesday and was complete by Friday morning (3:00 A.M.). Analysis of the wastewater after five hours of contact time showed the pH had dropped to 2.9, the TOC had been reduced by 74% and the COD had been reduced by greater than 70%. The relative toxicity of the untreated wastewater as measured by Microtox improved 10 fold (Figure 11) after treatment. As the Microtox EC_{50} increases the associated toxicity of the wastewater decreases. The color of the water had also improved from a dark brown to a very light green (almost clear). Eight days after the initial chemical dosing began, the wastewater was pumped into the primary clarifier of the wastewater treatment system.

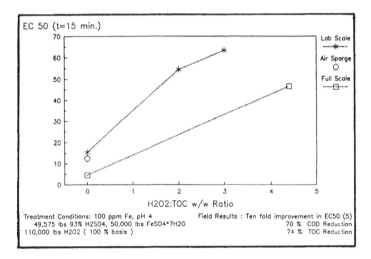

Figure 11. Relative Toxicity Reduction with H_2O_2
Pulp & Paper Wastewater Holding Pond

CHEMICAL PLANT CASE HISTORY

A southwestern chemical plant operated a diversion pond similar to the pulp and paper example previously described (Figure 12). The diversion pond of 225,000 gallons was also nearing capacity. The existing biological system was unable to oxidize the waste in the diversion holding pond. The wastewater had a COD of 1000 ppm and the treatment system was overloaded and close to its limit of effluent COD. The plant was currently hauling wastewater offsite for disposal to maintain an effective diversion capacity in their system. The cost of the offsite wastewater disposal was exceeding $30,000 per month. A large batch Fenton's type treatment was employed in an effort to make the waste more compatible with the existing wastewater treatment system. Although the reaction was similar to the previous pulp and paper example, dosing was based on COD instead of TOC. A 2:1 H_2O_2:COD w/w ratio was used with iron catalyzation (100 mg/L as Fe^{+2}).

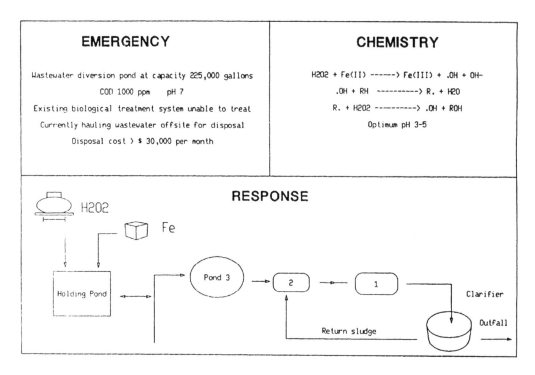

Figure 12. COD Removal From Chemical Plant Process Wastewater

The results after 24 hours showed a COD reduction of greater than 80% (Figure 13). The wastewater was then bled into the treatment system without any deleterious effect of the final effluent.

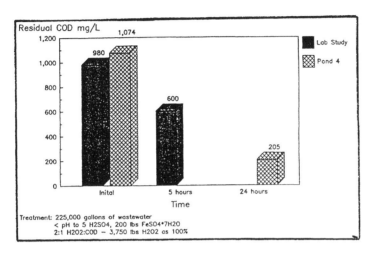

Figure 13. COD Reduction with Hydrogen Peroxide Chemical Plant Process Upset Wastewater

ACCIDENTAL SPILL RESPONSES

1.   TANKER ACCIDENT CASE HISTORY

A large tanker containing 900 tons of fermented molasses broke open and spilled it's contents into a marine dock along the River Mersey in England.[6] Approximately 3 million cubic meters of water were contaminated with this rapidly oxidizable substance. The effect was almost-immediate dissolved oxygen depletion in the dock area. The water turned black and began to evolve H_2S gas as sulfate-reducing bacteria began to proliferate. The water had a pH of 6.5-7, a DO of <0.3 and a H_2S concentration of 2-4 ppm. The authorities discussed several options for treating the waste; however, a hydrogen peroxide treatment proved to be cost effective. The hydrogen peroxide treatment proposed was based on reactions shown in Figure 14.

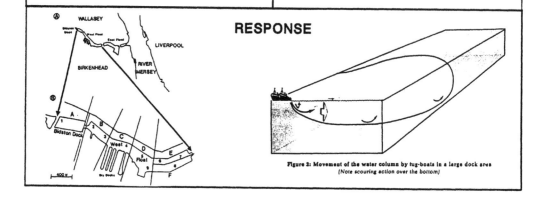

Figure 14. Sulfide Odor Removal From a Spill at a Marine Dock

A unique feature of the proposal was the employment of tug boats to mix the peroxide into the water column (Figure 14). The dock area was divided into 5 sections and each section was treated separately. The results of the treatment are shown in Figure 15. After 48 hours no residual peroxide was detectable using a colorimetric DPD field test kit. Normal boating and recreational activities were allowed to resume.

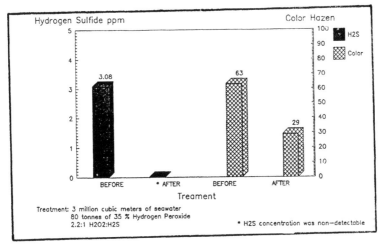

Figure 15.   Sulfide Control With Hydrogen Peroxide
Marine Dock Spill

2.   RAILROAD CAR LEAK CASE HISTORY

A railcar leak in a northeastern railroad switching yard resulted in a 17,000-gallon phenol spill. The contaminated soil was excavated and placed into 24 railcars while the contaminated railroad ties were placed in another 11 railcars. A treatability study was conducted to assess the effectiveness of Fenton's treatment on the soil (Figure 16).   The soil was placed in a

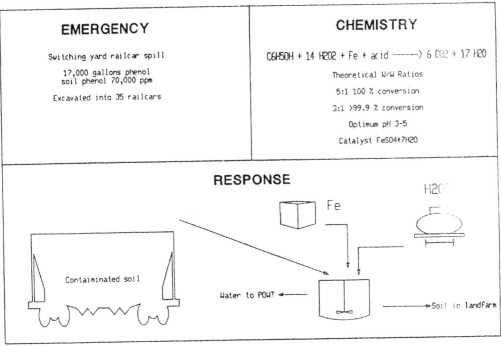

Figure 16.   Phenol Removal from Soil at a Railyard Spill

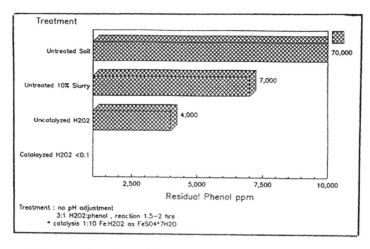

Figure 17. Phenol Oxidation in Soil With H_2O_2 Contaminated Soil from Railcar Spill

liquid-solid contactor as a 10% solids slurry. Analysis of the slurry, using a 4 AAP colorimetric field test kit, showed the phenol concentration to be 7000 ppm and the pH to be 7.4. The high level of phenol was expected to produce significant amounts of organic acid intermediates which would lower the pH to the optimum Fenton pH range of 3 to 5. Iron as $FeSO_4 \cdot 7H_2O$ was added to the slurry for catalyzation. A 3:1 H_2O_2:phenol w/w ratio was used with a reaction time of 1.5 to 2 hours. The results of the treatment are shown in Figure 17. The Fenton treatment of the slurry was shown to be capable of lowering the phenol concentration to levels where the water could be sent to a POTW and the soil could be landfarmed on site.

CONCLUSION

Hydrogen peroxide was chosen for these case histories because of its unique characteristics. The preceeding examples demonstrate hydrogen peroxide flexibility and versaltility in environmental emergency response situations. Hydrogen peroxide is an environmentally friendly oxidant which decomposes to water and oxygen. Of all the chemical oxidants available today, none can be applied as quickly and as easily as hydrogen peroxide delivered from isotainers.

SAFETY

It should be noted that hydrogen peroxide finds application in many industries; and, when used properly is a valuable and environmentally-safe chemical. However, as with most powerful chemicals, improper application or handling could create hazardous conditions or cause injuries to personnel. We strongly recommend you contact a manufacturer of hydrogen peroxide before experimenting with, designing, installing, or modifying an application system or using hydrogen peroxide.

66

REFERENCES

1. Cole CA, Stamberg JB and Bishop DF (1973). Hydrogen Peroxide Cures Filamentous Growth in Activated Sludge. EPA Report #67/2-73-033.

2. Brown RA and Norris RD (1982). Hydrogen Peroxide Reduces Sulfide Corrosion. Oil Sept. 6: 118-123.

3. Ayling GW and HM Castrantas (1981). Waste Treatment with Hydrogen Peroxide. Chem Engineering Vol 88(24): 79-82.

4. Bowers AR, Eckenfelder WW, Jr., Gaddipati P and Monsen RM (1987). Toxicity Reduction and Improvements in Biodegradability of Industrial Wastes Using Hydrogen Peroxide. Proceedings of the Oak Ridge Model Conference Vol 1(3): 153-163, Oak Ridge, Tenn.

5. Fraser JAL, Assoc. ed., "Landfill Leachate Treatment", Effluent and Water Treatment Jour, Oct. 87, p 49.

6. Fraser JAL, Assoc. ed., "Treatment of High Organic Waste", Effluent and Water Treatment Jour, Oct. 87, p 26.

WILFRIED EUL
GERD SCHERER
OSWALD HELMLING

Practical Applications of Hydrogen Peroxide for Wastewater Treatment

ABSTRACT

The increased environmental awareness requires reviewing production technologies and waste treatment processes differently. Previously disposed of waste has to be looked at again to verify whether its disposal meets current regulations.

The chemical destruction of waste products shows that it can solve past, present and future problems. Chemical treatments based on hydrogen peroxide (H_2O_2) or other active oxygen (AO) products (e.g. sodium percarbonate, sodium chlorate, etc.) are environmentally safe, effective and do not create hazardous byproducts.

Wastewater treatment systems based on H_2O_2 or AO products can be grouped into three (3) categories.

1. Municipal Wastewater Treatment

   H_2S abatement, peak or emergency oxygen supply, algae, slime and filamentous control.

2. Industrial Wastewater Treatment

   Destruction of cyanide, phenol, sulfide, sulfite, hypochlorite, formaldehyde, nitrite, etc.

3. Developmental Water Treatment

   Ozone - H_2O_2 peroxidation of drinking water, UV/H_2O_2 or UV/H_2O_2/O_3 for water detoxification and H_2O_2 for bioremediation. [13]

Examples will be given under each category in the paper and presentation.

Wilfried Eul, Degussa Corporation, 4 Pearl Court, Allendale, New Jersey, U.S.A.

Gerd Scherer, Degussa Corporation, 65 Challenger Road, Ridgefield Park, New Jersey, U.S.A.

Oswald Helmling, Degussa AG, Rodenbacher Chaussee 4, 6450 Hanau 11, Federal Republic of Germany

INTRODUCTION

Pollution can be defined as an undesired byproduct of human activities. Existing processes and technologies need to be reevaluated to determine whether changes can produce the same endproduct with less byproducts. Wherever the reduction of pollution is not possible, treatment technologies need to be implemented to render the generated byproducts harmless.

To totally eliminate pollution, one of the most advanced methods is oxidation of those materials. The end products of total oxidation are carbon dioxide (CO_2) and, more desirable, water (H_2O). A total oxidation is effectively achieved by biological treatment systems. Wherever the raw waste is toxic to the system, oxidative pretreatment using active oxygen chemicals renders biodegradable materials and harmless salts.

As an example, the destruction of phenols using Fenton's reagent (H_2O_2, low pH, catalyst) will render carboxylic acids which are easily biodegradable.

The group of oxygen and active oxygen based chemicals which are used for environmental purposes contain ozone (O_3), H_2O_2, O_2, sodium chlorite ($NaClO_2$), sodium percarbonate (SPC) and sodium persulfate (SPS), to name a few. This paper has its focus on H_2O_2 as an environmental chemical.

USES OF HYDROGEN PEROXIDE [12]

In addition to the use of H_2O_2 in markets such as pulp bleaching, textile bleaching, chemical synthesis, food additives and electronics, a substantial amount is used for environmental purposes. In this market segment, H_2O_2 finds its use to treat exhaust gases, industrial and municipal waste-waters and as an oxygen source for biological emergency situation. Within the area of wastewater treatment, H_2O_2 uses can be grouped into:

- municipal wastewater treatment
- industrial wastewater treatment
- evolving water and wastewater treatment

MUNICIPAL WASTEWATER TREATMENT

A classic use of H_2O_2 in the municipal wastewater treatment is H_2S abatement in addition to peak or emergency oxygen supply. The ever growing size of the municipal collecting systems has caused problems associated with odors, mostly hydrogen sulfide. The H_2S is generated by bacterial action and causes sever corrosion to sewer systems in addition to odor problems. The well known mechanism for the generation of H_2S and for the generation of sulfuric acid is illustrated on **Figure 1**.

Utilizing hydrogen peroxide, numerous installations throughout the country will be able to control the generated H_2S. The parameters related to an installation are listed in Table I. The objective of the use of hydrogen peroxide is odor abatement at the end of a force main.

A southern POTW experienced a severe odor problem and received numerous complaints from neighboring developments. On hot

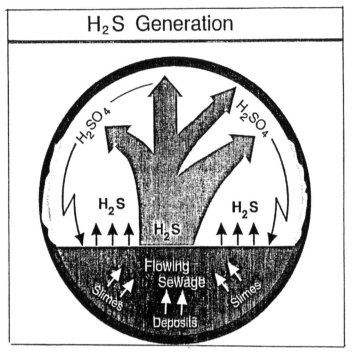

Figure 1 - Hydrogen Sulfide Generation in Flowing Services

Case Study		
Wastewater Flow	:	40 mgd
Average H_2S Content (May — Oct.)	:	8 ppm
pH (Average)	:	8.14
Average Use of H_2O_2 (August)	:	9.92 ppm
Mole Ratio (H_2S : H_2O_2)	:	1 : 1.24

TABLE I

summer days, when there was no air movement, we were able to detect an excess of 10 ppm H_2S (measured by Drager tubes) in the area of the grit chamber. This posed a substantial health risk for the workers at the plant. After reviewing the installation we were able to locate an injection point approximately 5 flow minutes up stream from the end of the force main. After installing a portable tank for hydrogen peroxide and connecting it to an appropriate dosing unit, we dosed the calculated amount of hydrogen peroxide continuously into the wastewater. Within less than 15 minutes a substantial reduction of the odor at the end of the pipe was detected. After several days of operation, the dose rate of hydrogen peroxide was fine tuned and the mass balance of H_2S vs. H_2O_2 could be calculated. The obtained mole ratio between H_2S and H_2O_2 was 1:1.24 which corresponds closely to the theoretically expected value.

Another area where hydrogen peroxide can be used in conjunction with biological treatment systems is for emergency oxygen supply. In Table II the parameters found at a POTW, where the high summer temperatures reduced the dissolved oxygen will below the bare minimum of 1 ppm are outlined. Hydrogen peroxide, which was brought on site as soon as possible, maintained a desirable dissolved oxygen system. The hydrogen peroxide was used until additional aeraters were installed which introduced air by mechanical means.

INDUSTRIAL WASTEWATER TREATMENT

Over the past 5-10 years, many uses for H_2O_2 in industrial wastewater treatment have been found. Due to expanding wastewater pretreatment programs, this trend will continue and H_2O_2 will be used increasingly. Some of the best known industrial applications for hydrogen peroxide are: [1,3,5]

- detoxification of cyanide, nitrite and hypochlorite
- destruction of phenol aromatics and formaldehyde
- removal of sulfite, thiosulfate and sulfide

From the above examples we will focus on the destruction of phenol, aromatics and hypochlorite. Table III gives an overview of a petrochemical process sidestream which contained an average of 8 ppm phenol. With the use of Fenton's reagent, the final phenol concentration was reduced to 8-12 ppb (GC-MS).

Another case was the destruction of aromatics, especially total phenol. Table IV gives an overview of another petrochemical process sidestream which was effectively treated with Fenton's reagent [7,10]. The primary goal in this application was to eliminate the toxicity of the water so that the subsequent biological treatment system could function properly. The total phenol content was consistently reduced to below 150 ppb, whereas the pure phenol content determined by GC-MS was below 2 ppb. A substantial reduction in the toxicity of the water was achieved.

Case Study

Wastewater Flow	:	11 mgd Daytime 17 mgd total
Dissolved Oxygen	:	0 – 0.15 ppm
H_2O_2 Usage (Daytime)	:	80 ppm
Dissolved Oxygen (After H_2O_2)	:	1.25 – 3.8 ppm

TABLE II

Case Study

Wastewater Flow	:	35 mgd
Phenol Content	:	Average 8 ppm
Fe^{2+} Usage	:	80 ppm
H_2O_2 Usage	:	17 ppm
Final Phenol Content	:	8 – 12 ppb

TABLE III

The destruction of hypochlorite was carried out in the depleted brine solution of a membrane cell chlor-alkali plant [2]. The removal of the hypochlorite is extremely important for this facility because any remaining hypochlorite will destroy the ion exchanger used in the water circle and will attack the membrane. The reaction between hypochlorite and H_2O_2 is extremely fast and, due to the significant ORP change, can be easily controlled. The automated H_2O_2 feed allows either a small amount of residual hypochlorite or a slight excess of H_2O_2. Table V outlines the parameters associated with this installation showing that we were well within the limits of the maximum allowable hypochlorite residue.

EVOLVING WATER AND WASTEWATER TREATMENT

Evolving technologies associated with hydrogen peroxide such as:

- UV/H_2O_2 in wastewater
- $UV/H_2O_2/O_3$ in wastewater [8,9]
- UV/O_3 in wastewater [11]
- H_2O_2/O_3 in water and wastewater [6]

will broaden the use of H_2O_2 in the 1990's.

The almost established advanced oxidation processes like UV/H_2O_2 and $UV/H_2O_2/O_3$ in wastewater enable the industry to break down some of the most stable compounds. Highly substituted aromatics, chlorinated aromatics, chlorinated aliphatics, PCBs, furans and dioxins can be destroyed using these technologies. These previously mentioned compounds could not, or only with substantial difficulty, be treated with Fenton's reagent.

The reason for the higher chemical activity is a generation of more or different radicals. **Figure 2** approximately shows the mechanism for the generation of those radicals.

Two case studies are presented from several installations where advanced oxidation technology is being used.

The first one was the result of a trial within the SITE Program of the EPA. The installation was used to destroy chlorinated organics, PCBs and pesticides. The initial concentration of up to 1 ppm was reduced to less than 10 ppb. The total treatment cost of $2.10/1,000 gallons includes the cost for generating the ozone, for buying hydrogen peroxide, the cost requirement for the UV lamps, the cost for the annual replacement of UV lamps, capital amortization and the estimated operation and maintenance cost. An overview of the parameters is shown on Table VI.

Table VII shows an overview of an installation which has been in operation since May 1989. The primary objective is the destruction of chlorinated solvents. The major constituent in the water, TCE, was successfully destroyed using $UV/H_2O_2/O_3$

Mechanism

$$H_2O_2 \xrightarrow{\text{UV-light}} 2 \; \cdot OH$$

$$H_2O_2 + O_3 \longrightarrow \cdot OH + \cdot OOH + O_2$$

$$O_3 \longrightarrow \cdot O + O_2$$

Figure 2 - Mechanism of UV - H_2O_2 Oxidation

Case Study

Wastewater Flow	:	12 mgd
Total Phenols	:	350 ppm
Phenol Content	:	85 – 100 ppm
Fe^{2+} Usage	:	150 ppm
H_2O_2 Usage	:	~500 ppm
Final Total Phenol	:	100 – 150 ppb
Final Phenol	:	< 2 ppb

TABLE IV

Case Study

Flow Rate	:	100 gpm
Pollutant Content	:	750 – 1000 ppb; VOC's, PCB's, Pesticides
Effluent Content	:	< 10 ppb
Total Treatment Cost	:	$2.11 / 1000 gal

TABLE VI

Case Study

Depleted Brine Flow	:	~580 gpm
Hypochlorite Content	:	120 – 270 ppm (Peak 2000 ppm)
H_2O_2 Usage (Month Avg.)	:	90 ppm
Residual Hypochlorite	:	0 – 6 ppm

TABLE V

Case Study		
Flow Rate	:	210 gpm
TCE Content	:	5.5 ppm
Effluent Content	:	~ 1 ppb TCE
Total Treatment Cost	:	$0.73 / 1000 gal

TABLE VII

technology. The TCE content was reduced from 5.5 ppm to approximately 1 ppb at a total treatment cost of $0.73/1,000 gallons. The total treatment cost again contains the generating cost for ozone, the price for hydrogen peroxide, the power and annual replacement for the UV lamps, operation and maintenance costs and capital amortization.

SUMMATION

Considering the above given case studies, hydrogen peroxide has its place as a chemical oxidant for pollution control. The ever growing municipal sewer systems and the need for emergency back-up for biological treatment systems will surely use hydrogen peroxide. The material also has its demand for industrial wastewater treatment. Expanding pretreatment programs will always demand innovative approaches where the versatility of hydrogen peroxide is applicable. The evolving technologies for water and wastewater treatment based on the advanced oxidation technologies are clear signs of this. In the future these (new) technologies will be state of the art and be listed within the regular industrial water treatment category. At this time hydrogen peroxide may get listed with the evolving treatment technologies to control sulfur dioxide and nitrogen oxides from exhaust gases [3]. In a few installations it has already been demonstrated that hydrogen peroxide can effectively treat a large variety of air pollutants.

Besides these direct uses of hydrogen peroxide for pollution control, there are certainly other industrial uses where the material renders cleaner effluents. H_2O_2 used in the bleaching of paper reduces the amount of chlorinated hydrocarbon in the effluent. H_2O_2 used as an oxidant in a large variety of chemical synthesis usually renders non-toxic byproducts which are easier to treat. With that in mind, you can see that the technologies based on hydrogen peroxide are a substantial contributor for a cleaner environment in the 1990's.

REFERENCES

[1]   Knorre H., Fischer J.:   Steels & Metals Magazine 1988 (Nr. 10), p 937-941.

[2]   Maennig D., Scherer G.; Proceedings of the US Chlorine Institute's 30th Plant Operations Seminar, Washington, DC, March 18, 1987.

[3]   Soldavini H., Wachendoerfer P., v. Wedel W.: VDI-Berichte (Germany) 1989 Nr. 730, p 331-347.

[4]   Keating E.J.:   Ind. Wat. Eng. 15 (1978) p 22.

[5]   Guittoneau S., de Laat J, Dore M., Duguet J. P., Bonnet C.:   Environmental Technology Letters 9 (1988), 1115-1128.

[6]   Glaze W.H., Kang J.-W.:   Research and Technology Journal AWWA 1988, May, p 57-63.

[7]   Fraser J.A.L., Sims A.F.E., Effluent and Water Treatment Journal 1984 (May), p 184-188.

[8]   Paillard H., Valentis G., Partington J., Tanghe H.:   Proceedings 1990 Spring Conf., Int'l. Ozone Assoc., Pan American Committee.

[9]   Peyton G.R., Le Farvre M.H., Gregory G.G., Fleck M.J.:   Proceedings 1990 Spring Conf., Int'l. Ozone Assoc., Pan American Committee.

[10]   Schumb W.C., Satterfield C.N., Wentworth R.L.: Hydrogen Peroxide, 1955 New York, Reinhold Publ.

[11]   Clarke N., Knowles G.:   Effluent and Water Treatment Journal 1982 Nr. 9, p 335-340.

[12]   Ullmann's Enzyklopädie der Technischen Chemie, 4. Aufl.   1979 Vol. 17, Weinheim-Heidelberg-New York, VCH - Publishers, p 691-718.

[13]   Hinchee R.E., Downy D.C.:   NWWA/API Proc. Conf. & Exhib.   "Petroleum Hydrocarbons and Organic Chemicals in Groundwater," Houston, TX, Nov. 1988.

MALCOLM BUCK
JENNIFER CLUCAS
COLIN McDONOGH
STEPHEN WOODS

NO$_x$ Removal in the Stainless Steel Pickling Industry with Hydrogen Peroxide

ABSTRACT

Stainless steel pickling is a major generator of NOx. Hydrogen peroxide technologies have been developed, and proven, to remove or suppress emissions with no change to the pickling process. Case histories are presented which demonstrate the efficacy for removal, by post-pickling gas scrubbing, and suppression, by addition to the pickling bath.

INTRODUCTION

The increasing enactment of legislation to improve the quality of industrial gas discharges has resulted in the need for proven, cost effective methods of controlling air pollution. This is heightened by a growing public appreciation of the dangers of airborne pollution. An area of special concern is the emission of NOx gases.

The major constituents of NOx are Nitric Oxide (NO) and Nitrogen Dioxide (NO$_2$), though small quantities of Dinitrogen Tetroxide (N$_2$O$_4$) and Dinitrogen Trioxide (N$_2$O$_3$) may be present, due to the equilibrium reaction between NO and NO$_2$. Both NO and NO$_2$ are classified as poisons and are toxic to humans via inhalation. NOx can also be directly responsible for plant damage and, indirectly, via its oxidation, contributes to acid rain concerns. Since NO$_2$ is a red-brown coloured gas, NOx emissions are readily identifiable. Conversely, the absence of a brown plume, from a potential NOx emitting source, is a qualitative measure of NOx suppression.

Metal surface treatments and particularly stainless steel pickling, are major generators of NOx. A typical pickling process will employ a liquor comprising 15% Nitric acid and 3% Hydrofluoric acid. It is the reaction between the Nitric acid and transition metals which is responsible for the NOx generation. A number of methods to control the emission of these gases have been proposed and these are summarised in Table I.

Malcolm Buck, Jennifer Clucas, Colin McDonogh and Stephen Woods, Interox Research and Development, Widnes, Cheshire, England

NOx suppression in steel pickling

TABLE 1

Treatment	Catalytic reduction by NH_3	Gas Scrubbing by NaOH Solution	Gas Scrubbing by H_2O_2	Addn.to bath H_2O_2	$CO(NH_2)_2$
Capital Cost	very high	high	high	low	low
Variable Cost	low	low	high	high	low
HNO_3 Consumption	no influence	no influence	lower	lower	higher
NO_x Reduction	very high	high	very high	very high	low
By-products	no problem	difficult to discharge	recovered	recovered	no problem
Steel Surface Quality	no influence	no influence	no influence	better	worse

Catalytic reduction with ammonia, whilst being extremely effective, is very expensive in terms of capital cost. The use of urea has been proposed but this necessitates a higher nitric acid consumption and results in a poorer stainless steel surface quality. For these reasons, the method of control most frequently adopted is gas scrubbing, using an alkaline liquor of sodium hydroxide. This method is effective and does not impair the quality of stainless steel. However, whilst the capital cost is lower than for ammonia reduction, it is still relatively high. Further, the process results in the formation of large quantities of sodium nitrate which must be disposed of, with an associated cost.

Two methods have been developed to minimise NOx emissions using Hydrogen Peroxide (H_2O_2). This may be added to either the gas scrubbing liquor or directly to the pickling bath.[1] Both techniques are extremely effective at controlling NOx and have no detrimental effect on the stainless steel quality. Indeed addition to the pickling bath may even improve the surface quality. Additionally the consumption of nitric acid in the pickling process is reduced. Both methods have been proven on a commercial scale.

REACTIONS BETWEEN H_2O_2 AND NOx

The essential chemistry of both processes is identical, since the reaction between H_2O_2 and NOx occurs in the aqueous phase. For the addition to the bath this is in pickling liquor itself, whilst for gas scrubbing the reaction follows absorption of the NOx gases into the aqueous scrubbing liquor.

TABLE II

CHEMISTRY OF NOx ABSORPTION AND REACTION WITH HYDROGEN PEROXIDE

(A) $\underline{NO_2}$

$$2NO_2 \text{ (G)} \rightleftharpoons N_2O_4 \text{ (G)} \longrightarrow N_2O_4 \text{ (AG)}$$

$$N_2O_4 \text{ (AQ)} + H_2O \longrightarrow HNO_2 + H^+ + NO_3^-$$

$$HNO_2 + H_2O_2 \longrightarrow HNO_3 + H_2O$$

$$2NO_2 + H_2O_2 \longrightarrow 2HNO_3$$

(B) $\underline{NO/NO_2}$

$$NO + NO_2 \rightleftharpoons N_2O_3 \text{ (G)} \longrightarrow N_2O_3 \text{ (AQ)}$$

$$N_2O_3 \text{ (AQ)} + H_2O \longrightarrow 2HNO_2$$

$$2HNO_2 + 2H_2O_2 \longrightarrow 2HNO_3 + 2H_2O$$

$$NO + NO_2 + 2H_2O_2 \longrightarrow 2HNO_3 + H_2O$$

(C) $\underline{NO}$

$$NO \text{ (G)} \longrightarrow NO \text{ (AQ)}$$

$$NO \text{ (AQ)} + H_2O_2 \longrightarrow NO_2 \text{ (AQ)} + H_2O$$

$$3NO_2 + H_2O \longrightarrow 2HNO_3 + NO$$

$$2NO + 3H_2O_2 \longrightarrow 2HNO_3 + 2H_2O$$

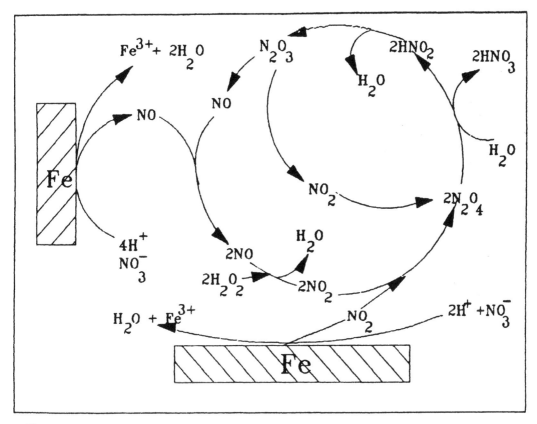

Figure 1

Stainless steel pickling – NOX Cycle

In the aqueous phase NOx reacts with water to form Nitrous acid (HNO_2). The HNO_2 is relatively unstable and will readily decompose back to NO_2, NO and H_2O. These NOx species would ultimately be emitted from the process. However, the presence of H_2O_2 rapidly oxidises the HNO_2 to the more stable HNO_3, thus preventing the reformation and emission of NOx.

These reactions are summarised in Table II, and shown schematically in Figure 1 as an NOx cycle.

NOx REMOVAL BY GAS SCRUBBING WITH H_2O_2

There are a number of techniques available for contacting gas and liquid phases, in order to remove,or "scrub" pollutants from the gas phase by mass transfer to the liquid phase. This leads to a variety of equipment types. Perhaps the most important and widely used are packed towers.

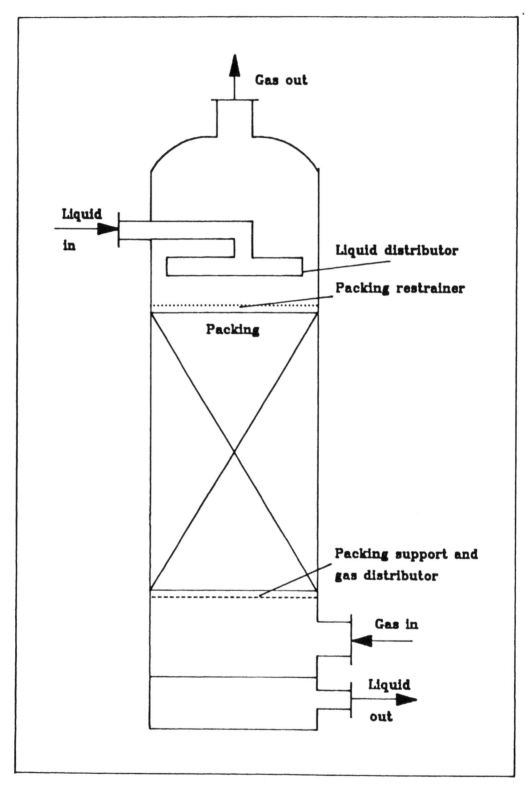

Figure 2. Schematic Diagram of a Packed Column

Gas scrubbers of this type usually effect the contact of the two phases by random packings, such as rings. The tower diameter is fixed by the mass flowrates of both the gas and the liquid, and the height is varied to achieve the desired pollutant removal.

Gas scrubbers generally operate counter-currently with the scrubbing liquor distributed over the packing from the top of the tower. In the case of H_2O_2 assisted gas scrubbing this liquor consists of HNO_3 and H_2O_2, at typical concentrations of 20% w/w and 0.5% w/w respectively. The gas enters the tower below the level of the packing and contacts the liquid as it passes upward through the tower, thereby transferring NOx from the gas to the liquid phase. This is shown schematically in Figure 2.

Interox have developed a number of computer programmes which utilise the above chemistry and technology to design gas scrubbing units for the removal of NOx. These programmes were used to design a scrubber for a European stainless steel mill. The requirements were that it should treat

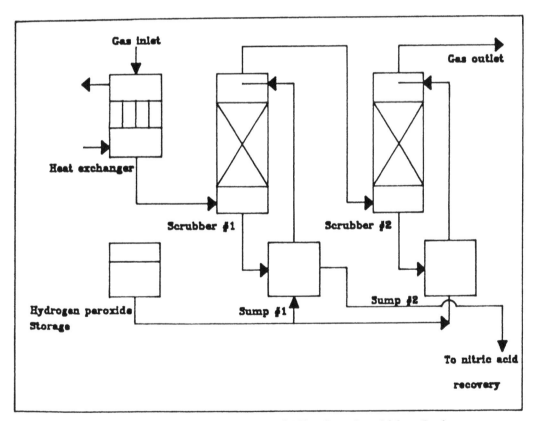

Figure 3. Schematic Diagram of NOx Gas Scrubbing System

TABLE III

	Inlet	Scrubber #1 Outlet	Scrubber #2 Outlet
NOx Conc./ppm	1500–2000	400–600	50
NO Conc./ppm	700–1200	150–200	30
Gas Temperature/degC	45–50	–	30–35

a gas flow of 8-9000 $m^3 h^{-1}$ containing 1500-2000 ppm NOx, to produce an outlet concentration of less than 50 ppm, this discharge limit being set by local authorities. Several other techniques were considered, but none could guarantee to achieve the targets.

The Interox design incorporated a two stage scrubbing process with H_2O_2 injected into the scrubbing liquor of both towers. This design is shown schematically in Figure 3. The performance of the gas scrubbing unit is shown in Table III, showing clearly the design targets had been achieved.

An additional benefit of the use of H_2O_2 is that the by-product of the scrubbing process is Nitric acid, instead of sodium nitrate when using sodium hydroxide. This acid is recovered in usable concentrations and recycled back to the pickling process. Thus not only is the sodium nitrate waste disposal cost eliminated but the overall nitric acid consumption is reduced.

This plant has been operating successfully since March 1988. The H_2O_2 consumption in that period is only slightly in excess of the stoichiometric requirement, indicating the high efficiency of the design.

NOx SUPPRESSION BY H_2O_2 INJECTION TO THE PICKLING BATCH

The key to the efficient usage of H_2O_2 in NOx suppression, by addition to the pickling bath, lies in effective mixing. When it is added to the pickling liquor, containing both oxides of nitrogen and transition metal ions, it will either oxidise the NOx, according to the chemistry outlined in Table II, or will itself undergo catalytic decomposition by reaction with the metal ions. The oxidation reaction is marginally faster than the decomposition reaction and thus, with efficient mixing, will proceed preferentially.

One method of effective mixing of H_2O_2 and the bath liquor is to recirculate the liquor. This technique has been patented, though only in Sweden, by Eka Nobel.[2] The bath contents are pumped around a recirculation loop at a rate of 5-10 bath changes per hour. H_2O_2 (35%) is dosed into this loop at 1-2 litre per minute. A schematic diagram is shown in Figure 4. NOx suppressions in excess of 90% have been achieved by this technique.

However the installation of a recirculation loop can be high in terms of capital cost. An alternative, developed by Interox, is a system which

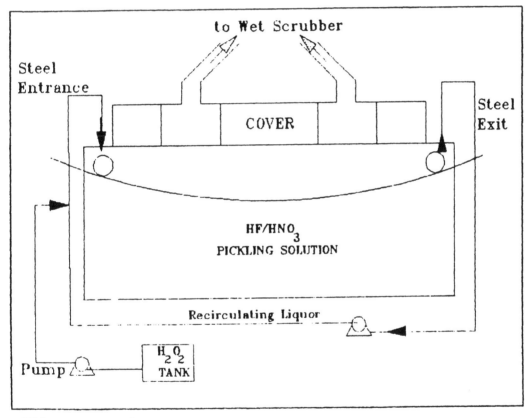

Figure 4

NOX abatement with hydrogen peroxide
addition via recirculation loop

injects the H_2O_2 directly into the pickling bath via a bifurcated sparge, located at the bottom of the tank. This is shown in Figure 5.

This system has been proven in a number of trials. Measurements taken show that there is a distinct relationship between the amount of NOx emitted from the bath and the concentration of nitrites in the bath liquor. These concentrations are also related to the speed at which the stainless steel is passed through the bath. This relationship continues with the amount of H_2O_2 added to the bath. Thus increasing the H_2O_2 addition, at constant line speed, reduces the nitrite in solution and thereby NOx emission. These features are illustrated in Figures 6 to 9. As a consequence of these relationships, if H_2O_2 addition is increased, as the line speed is increased, there should be no change in the amount of NOx evolved. This may provide a convenient method for the control of H_2O_2 addition, ie. linking the H_2O_2 dosage rate to speed of the line.

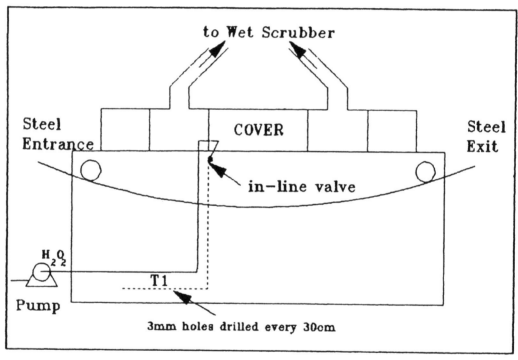

Figure 5

NOX abatement with H_2O_2 addition directly to the pickling bath

NOx abatement with hydrogen peroxide

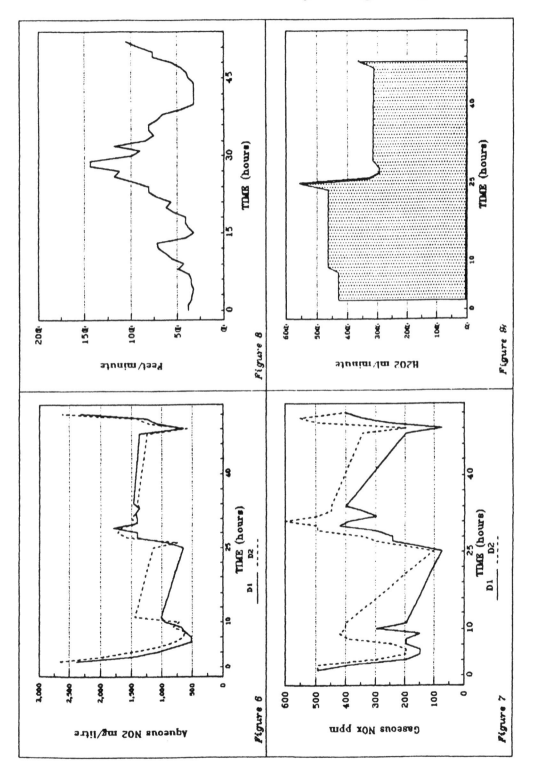

Figure 6

Figure 7

Figure 8

Figure 8:

The results of these trials have been very encouraging and the process is now being used on a commercial scale. On this scale it has been found that the reduction in HNO_3 consumption partially offsets the cost of H_2O_2. Further the quality of the stainless steel surface is improved.

SUMMARY

We have demonstrated how H_2O_2 can be used effectively, in two ways, to prevent NOx emission from a stainless steel pickling process. It has the advantage over other control methods in that the by-product of the reaction, HNO_3, can be recycled back into the pickling process, thereby achieving a raw material saving. Further, when added to the pickling bath, it may also result in an improved surface quality of stainless steel.

In common with other effluent treatments it is easier to reduce concentrations from several thousand ppm to tens of ppm, than from hundreds to tens of ppm. For this reason the two H_2O_2 based systems are rarely run in tandem, ie. using gas scrubbing as a polishing stage. The choice of method is therefore dictated by the individual plant situation. A plant with an existing scrubbing system can be modified to operate with H_2O_2 at relatively low capital cost. However if no scrubber system is in place these costs are likely to be high, and direct injection into the pickling bath would be the preferred option.

Hydrogen peroxide, when used correctly, is an extremely efficient and safe method of resolving NOx emission problems. Interox personnel would be pleased to advise on all aspects of the use and handling of H_2O_2.

Finally, examination of the problems of NOx evolution, from HNO_3 based liquors, has led to a consideration of the role of this acid. Nitric acid provides both an acid and an oxidant to the process. These functions could be replaced, at least theoretically, by sulphuric acid and H_2O_2 respectively.

Tests, both in the laboratory and on plants in Europe, up to a 20000 litre scale, have shown that, with a suitable stabiliser, a sulphuric acid/hydrogen peroxide/hydrofluoric acid pickling liquor is just as effective as the conventional nitric/hydrofluoric acid mixture. Thus H_2O_2 technology holds out to the stainless steel industry the ultimate solution to NOx emission - a nitric acid free process.

REFERENCES

(1) US PATENT 3,945,865
(2) EUROPEAN PATENT 259,533

RAYMOND C. LINNEMAN
T. HOUSTON FLIPPIN

Hydrogen Peroxide Pretreatment of Inhibitory Wastestream—Bench-Scale Treatability Testing to Full-Scale Implementation: A Case History

ABSTRACT

The BF Goodrich Henry Plant discharged a wastestream which caused activated sludge inhibition at flow rates as low as 5 gpm. To maintain wastewater treatment facility compliance, BF Goodrich contract hauled and disposed of that portion of the wastestream which could not be processed by the WWTF. Costs of this operation prompted a wastestream pretreatment evaluation.

Batch treatability tests were used to screen the effects of alternative pretreatment technologies on rendering this wastestream non-inhibitory and biodegradable. This batch testing revealed that thiosulfate reduction and hydrogen peroxide oxidation warranted further investigation. Following bench-scale continuous flow testing, hydrogen peroxide pretreatment was selected as the treatment technology. A process design was generated and a full scale pretreatment facility was constructed.

The full-scale facility has been successfully processing all of this pretreated wastestream (20 gpm) for eight months. Activated sludge response to this pretreated wastestream correlated well with that predicted by the bench-scale tests.

BACKGROUND

The BF Goodrich Company owns and operates a chemical manufacturing plant in Henry, Illinois. Associated with this plant is an on-site activated sludge wastewater treatment facility (WWTF). Wastewaters discharged to the WWTF vary depending on production campaigns. One wastestream (W-18), was observed to cause activated sludge inhibition and effluent BOD violations at discharge rates as low as 5 gpm. However, during the associated production campaign, the discharge flowrate of W-18 was

Raymond C. Linneman, Utilities Manager, BFGoodrich Company, Henry, Illinois, U.S.A.
T. Houston Flippin, P.E., Project Manager, Eckenfelder Inc., Nashville, Tennessee, U.S.A.

55 gpm. BF Goodrich collected all of the W-18 not discharged to the WWTF and transported this wastestream 100 miles for off-site disposal. This method of wastewater disposal was very costly.

TREATABILITY EVALUATION

A two phased treatability study was initiated: batch screening tests for selection of the two most cost-effective methodologies and bench-scale continuous flow testing of these two selected pretreatment technologies.

BATCH SCREENING TESTS

Aliquots of W-18 were subjected to several pretreatment technology screening tests. The pretreated samples were then subjected to Fed Batch Reactor (FBR) activated sludge acute biodegradation tests. A summary of the results is presented in Table I.

Hydrogen peroxide oxidation and sodium thiosulfate reduction were selected as the two most cost-effective pretreatment technologies which improved W-18 biodegradability.

TABLE I - W-18 TREATABILITY SCREENING TEST RESULTS

PRETREATMENT	SUBSTRATE REMOVAL RATE (mg TOC/g VSS•hr)
No Pretreatment	18.0
Acidification - pH 1.5, 100°C, 1 hr	19.8
Alkalinization - pH 12, 100°C, 1 hr	33.3
Solvent Extraction	29.6
Evaporation	17.8
Carbon Treatment	
• 3 g/l	47.3
• 10/g/l	57.9
Permanganate Oxidation (5.0 g/l $KMnO_4$)	42.2
Sodium Thiosulfate Reduction (1.5 g/l $Na_2S_2O_3$)	41.4
Peroxide Oxidation (5 g/l H_2O_2)	55.6

Hydrogen peroxide pretreatment consisted of lowering W-18 pH to 3.5 with concentrated sulfuric acid, addition of 5,000 mg/l H_2O_2 and 915 mg/l Fe^{++}, contacting for two hours, elevating to pH 7.0, and discharging to activated sludge system. Sodium thiosulfate pretreatment consisted of addition of 1,500 mg/l $Na_2S_2O_3$ and providing one hour contact time. No benefit was observed at higher dosages of hydrogen peroxide or sodium thiosulfate. Both pretreatment processes were followed by primary sedimentation.

W-18 was pretreated using hydrogen peroxide and sodium thiosulfate as described above. Following this pretreatment, pretreated W-18 was progressively added to the total WWTF influent and fed to 20-ℓ activated sludge treatability units operating at a 25 day mean cell residence time. A parallel treatability unit (the Control) receiving no W-18 was operated at the same MCRT. Pretreated W-18 feed rates were increased initially from 10 gpm to 30 gpm over the course of 5 weeks. Even with this acclimation period, the treatability unit receiving sodium thiosulfate pretreatment could not tolerate the desired 30 gpm loading rate. The treatability unit receiving hydrogen peroxide pretreatment was capable of handling the desired 30 gpm loading rate (Figures 1 and 2).

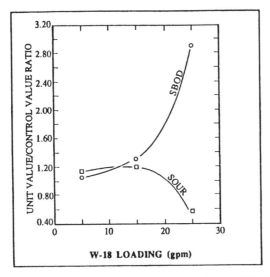

Figure 1. Effect Of W-18 Loading On Effluent Soluble BOD And Aeration Basin SOUR For Unit Receiving Sodium Thiosulfate Pretreatment

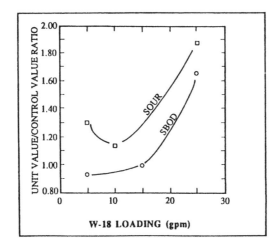

Figure 2. Effect Of W-18 Loading On Effluent Soluble BOD And Aeration Basin SOUR For Unit Receiving Hydrogen Peroxide Pretreatment

Once hydrogen peroxide pretreatment was selected as the preferred pretreatment process, several pretreatment alternatives were evaluated. These alternatives are presented below:

- 5,000 mg/l H_2O_2, 915 mg/l Fe^{++}, pH 3.5, 120 minute contact time and primary sedimentation.

- 3,500 mg/l H_2O_2, 1,154 mg/l Fe^{++}, pH 2.8, 30 minute contact time and filtration.

- 1,700 mg/l H_2O_2, 320 mg/l Fe^{++}, pH 2.8, 30 minute contact time and sedimentation.

Results from this alternative evaluation indicated that effluent compliance (i.e., 17 mg/l SBOD) could be achieved without effluent filtration and at reduced hydrogen peroxide dosages (i.e. < 5,000 mg/l) and contact times (i.e., 30 minutes). It was recommended that the pretreatment system be designed for the more conservative operating conditions (5,000 mg/l H_2O_2, 1,154 mg/l Fe^{++}, pH 2.5 to 3.5, and 120 minute contact time). System optimization would be determined through field experience.

FULL-SCALE SYSTEM DESIGN AND OPERATION

The pretreatment system installed based on the bench-scale treatability evaluation consists of two (2) agitated FRP tanks, a ferrous sulfate addition system, a hydrogen peroxide addition system and pH control systems (See Figure 3). The wastewater from the process flows into the first tank (Tank 1) at 50 gpm where a continuous feed of 0.9 gpm of 25 percent ferrous sulfate is added. The pH in this tank is controlled at 3.0 by the addition of sulfuric acid. The level of Tank 1 remains reasonably constant as the contents are pumped to Tank 2. In this tank 49 percent hydrogen peroxide is added at 0.6 gpm. The pH of Tank 2 discharge is monitored and adjusted up to 8.0 with caustic prior to being pumped to a 400,000 gallon storage tank. The W-18 process normally operates 10 days per month therefore the storage tank is used to equalize the flow of this wastestream to the WWTF.

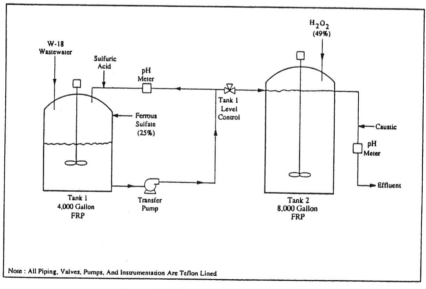

Figure 3. W-18 Wastewater Pretreatment System

TABLE II - EFFECTS OF ALTERNATIVE HYDROGEN PEROXIDE PRETREATMENT METHODOLOGIES

Pretreatment Process	Effluent SBOD (mg/l)	Aeration Basin	
		SOUR (mg O_2/g VSS•hr)	F/M (g SCOD/g VSS•day)
5,000 mg/l H_2O_2, 915 mg/l Fe^{++} pH 3.5, 120 minute and primary sedimentation	9	12.5	0.53
1,750 mg/l H_2O_2, 320 mg/l Fe^{++}, pH 2.8, 30 minute and primary sedimentation	25	13.4	0.47
3,500 mg/l H_2O_2, 1,154 mg/l Fe^{++} pH 2.8, 30 minute and filtration	8	10.3	0.53

Operationally the pretreatment system performs extremely well. Ferrous sulfate and hydrogen peroxide additions present no problems and pH control is reliable. Some solids are generated as a result of the ferrous sulfate addition; however, agitation in both tanks is sufficient to prevent any pluggage throughout the system.

Approximately 6 months after the pretreatment system was started up some concerns developed that potentially high levels of an organic hydrocarbon were being liberated from the wastewater as a result of pretreatment. Further laboratory tests showed that by dropping the addition of the ferrous sulfate from the pretreatment scheme, virtually none of this organic hydrocarbon was liberated. However, the COD reduction of the wastewater was also reduced (See Table III). In order to obtain the same COD reduction without using ferrous sulfate addition the hydrogen peroxide addition rate was doubled. Operationally, the pretreatment system remained the same. The pH of the effluent was still adjusted to 8.0.

TABLE III - W-18 WASTEWATER ORGANIC CONCENTRATIONS

	SCOD (mg/l)	SBOD (mg/l)	STOC (mg/l)
Raw Process Wastewater	22,000	1940	7,500
Pretreatment Effluent	11,970	--	4,000
Pretreatment Effluent (H_2O_2 only)	13,300	1200	4,150

FULL-SCALE WWTF PERFORMANCE

Prior to start up of the pretreatment system, the maximum feed rate of W-18 wastewater to the waste treatment system was 8 gpm. Increasing the feed of this waste above this amount would result in reduced activity of the WWTF biomass. The pretreatment system as initially designed with ferrous sulfate addition, allowed for the feed rate to be on the average at 18 gpm with a peak of 23 gpm without any noticeable impact of the system. Since ferrous sulfate addition has been eliminated from the pretreatment system, there has been a slight increase in the pretreated W-18 COD. Feed rates to the WWTF of hydrogen peroxide treated W-18 have only been recently 13 gpm on the average with a peak of 16 gpm. These flow rates have had no adverse affect on the biological system (see Table IV). The WWTF has historically exhibitted SOURs of 6.0 to 9.0 mg/g•hr and effluent TBOD concentrations of 6 to 14 mg/l without the presence of W-18 wastewater. Due to aeration system limitations higher feed rates of W-18 wastewater have not been possible.

The primary success of the W-18 pretreatment system has been to reduce the inhibitory affect of the wastewater on the WWTF biomass. In addition, pretreatment of W-18 wastewaters significantly reduced WWTF influent BOD loading.

In recent months, pretreated W-18 discharge rates have been limited to 16 gpm since addition of ferrous sulfate was eliminated, but further treatability testing, not presented in this paper, indicates that feed rates as high as 30 gpm will not adversely affect the WWTF biomass.

TABLE IV - EFFECTS OF PRETREATED W-18 LOADING ON THE FULL-SCALE WWTF

	Feed Rate (gpm)	SOUR (mg/g/hr)	Effluent TBOD (mg/l)	Effluent TCOD (mg/l)
Pretreated W-18	12.4	8.68	7.1	226
	15.4	6.81	6.8	238
	16.8	7.58	7.5	234
	18.0	6.53	12.0	242
Pretreated W-18 (H_2O_2 only)	8.0	7.24	8.6	244
	9.0	6.74	13.6	447
	12.0	8.18	7.4	215
	13.0	7.48	6.3	179

SUMMARY AND CONCLUSIONS

Without hydrogen peroxide pretreatment, discharge of W-18 wastewaters to the on-site WWTF was limited to 5 to 8 gpm due to activated sludge inhibition and associated BOD permit violations. With hydrogen peroxide pretreatment, W-18 wastewaters can be discharged at rates of 30 gpm to the on-site WWTF without activated sludge inhibition and within effluent

BOD compliance. No further off-site disposal of W-18 wastewaters has been required since initiation of hydrogen peroxide pretreatment, a substantial savings.

Careful consideration should be given to off-gas quality when designing a hydrogen peroxide pretreatment system (especially when using Fenton's chemistry). Additional care should be taken in the selection of compatible materials of construction.

Cyanide Detoxification with Peroxygens

ABSTRACT

The common process of oxidizing cyanide with hypochlorite often results in the formation of chlorinated compounds (AOX). We present an alternative process for the detoxification of cyanide using hydrogen peroxide (H_2O_2) and/or potassium monopersulfate (CUROX^R). A monitoring system to control reaction is described and examples are presented. Comparisons regarding the AOX-formation (AOX-de novo) are given.

REVIEW OF COMPETITIVE PROCESSES

Prussic acid and their salts (cyanides) are well known as highly toxic. Therefore cyanide-containing waste waters have to be treated according to the laws of FRG, whether fed into sewers or into receiving bodies of water. The maximum concentration allowed after treatment is 0,1 mg/l.

In the FRG such a treatment must be in accordance to the "common rules of technology". This means that for the treatment a process has to be applied which is well known and (more important) which is verified in a number of existing plants on a technical scale.

In the past, the process mainly used [1],[2],[3] was the treatment using hypochlorite (NaOCl). The reaction is simple and effective and can be controlled and therefore automated by the redox potential:

$$CN^- + ClO^- + H_2O \longrightarrow ClCN + 2OH^- \qquad (1)$$

$$ClCN + 2OH^- \longrightarrow CNO^- + Cl^- + H_2O \qquad (2)$$

Dr. Karl Heinz Gregor, Peroxid-Chemie GmbH, D-8023 Höllriegelskreuth, Federal Republic of Germany

followed by (depending on reaction conditions)

$$CNO^- + 2H_2O + 2H^+ \longrightarrow NH_4^+ + CO_2 + H_2O \qquad (3)$$

or

$$2CNO^- + 3ClO^- + 2H^+ \longrightarrow N_2 + 2CO_2 + 3Cl^- + H_2O \qquad (4)$$

The reactions (1) to (4), however, do also include side reactions. Although the oxidation is rather selective, a number of organic compounds in the waste water can be chlorinated, which results in the formation of AOX (halogenated hydrocarbons adsorbable on activated carbon), which is called "de novo-AOX". The amount of de novo-AOX depends on sort and quantity of organic compounds in the waste water as well as on how easily these compounds can be chlorinated under alkaline conditions. In addition, it is possible that organic pollutants are split off into fragments by oxidation and will then form AOX, e.g. trichloromethane.

Alternative treatment methods have been offered in the past to perform the oxidation of cyanides by applying ozone or electrical energy [4],[5].

The reaction with ozone is

$$CN^- + O_3 \longrightarrow CNO^- + O_2 \qquad (5)$$

$$CNO^- + 2H^+ + H_2O \longrightarrow CO_2 + NH_4^+ \qquad (6)$$

or

$$2CNO^- + H_2O + 3O_3 \longrightarrow 2HCO_3^- + 3O_2 + N_2 \qquad (7)$$

with side reactions

$$CNO^- + NH_4^+ \longrightarrow H_2NCONH_2 \qquad (8)$$

$$H_2NCONH_2 + O_3 \longrightarrow N_2 + CO_2 + 2H_2O. \qquad (9)$$

The electrochemical oxidation can be described with

$$CN^- + 2OH^- \longrightarrow CNO^- + H_2O + 2e^- \qquad (10)$$

$$CNO^- + 2H^+ + H_2O \longrightarrow NH_4^+ + CO_2 \qquad (11)$$

or

$$2CNO^- + 4OH^- \longrightarrow 2CO_2 + N_2 + 2H_2O + 6e^- \qquad (12)$$

Neither method shows high selectivity, so not only cyanide, but also a number of different pollutants are oxidized. This results in high energy costs, a severe disadvantage against other methods.

Other proposals to detoxify cyanides have been mentioned [1],[3] for special applications, e.g. precipitation in form of "Berliner Weiß" or "Berliner Blau", also known as "Turnbulls Blue", a compound of hexacyano complexes which contains Fe(II) as well as Fe(III). These reactions result in solid waste, creating problems for further treatment and disposal. Our opinion is that we should rather avoid such problematic waste than to produce it needlessly.

CYANIDE DETOXIFICATION WITH HYDROGEN PEROXIDE

The detoxification process with hydrogen peroxide has been well known for a number of years. At least one German company is known to have used hydrogen peroxide since 1975 (later completed by using CUROX[R] for detoxifying very low cyanide concentrations).

The most important advantage of hydrogen peroxide compared to other chemicals is that it decomposes into water and oxygen after reaction. Thus the salt concentration in the treated waste water is not increased and the reaction is really an oxidizing instead of a chlorinating one. Thus the formation of AOX is suppressed and during our investigations a reduction of AOX could even be demonstrated.

The reaction with hydrogen peroxide is [1],[2],[3],[6]

$$CN^- + H_2O_2 \longrightarrow CNO^- + H_2O \qquad\qquad (13)$$

$$CNO^- + 2H_2O + 2H^+ \longrightarrow CO_2 + NH_4^+ + H_2O. \qquad\qquad (14)$$

The only problem to be solved is that hydrogen peroxide does not show a concentration-dependent redox potential. In other words - the redox potential of an aqueous solution does not change with addition or removal of hydrogen peroxide, with the exception that it reacts with other compounds showing a redox potential. One of these compounds is cyanide. Down to very low concentrations cyanides are forming a well defined potential (silver/calomel system) which can be used to control the oxidizing reaction with hydrogen peroxide. In Germany the company PROMINENT DOSIERTECHNIK, Heidelberg, which produce measurement instruments and control systems (also lab equipment, pumps, etc.) developed a measuring and controlling device - Dulcotrol-Detox - with the assistance of PEROXID-CHEMIE GMBH. This device enables the automatization of the oxidation reaction between cyanide and hydrogen peroxide. A typical recording of the proceedings can be seen in TABLE I. The original cyanide concentration of 1.310 ppm could be reduced under alkaline conditions with hydrogen peroxide down to 1,1 ppm, while the electrochemical potential is jumping from -650 mV to approximately 0 mV. The effectiveness of the reaction is 99,9 %, however, the limit of 0,1 ppm, could not be reached. We do not know exactly why this interruption of the oxidation occurs, it may be due to complexes in the aqueous solution. Nevertheless, we are able to offer methods to avoid this by either adding an activating chemical compound (activator HA) or the potassium salt of Caro's acid, which we call "CUROX[R]". After adding 1 g/l of CUROX[R] to the solution, the cyanide concentration of 1,1 ppm drops down to a non-detectable limit, although no change in the electrochemical potential like during addition of hydrogen peroxide can be seen.

The amount of 1 g/l of CUROX[R] results from experience in lab scale trials. Our recommendation for running the process in technical scale is to reduce the dosage as far as possible in order to achieve an optimized detoxification from both the technical and economical point of view.

TABLE I: TYPICAL ISSUE OF A CYANIDE DETOXIFICATION

time	temperature	potential	CN⁻—concentration	remarks
(min)	(°C)	(mV)	(mg/l)	
0	22,5	−650	1,310	pH=9,7
15	24,5	−530		start H_2O_2 dosage
30	25,8	− 7		dosage slows down
33	27,0	±0		potential plateau
35	28,2	+20		colour change to green
45	28,7	+11	1,1	photometrical analysis
55	29,0	+7		addition of 1 g/l Curox
85	28,3	+3	n.d.	end of reaction
n.d. = not detectable				

To perform the reaction with hydrogen peroxide our recommendations are:

- concentration of cyanide should not be higher than approximately 5-10 g/l; otherwise the reaction might be too intensive and then could not be kept under control

- the pH should be adjusted to approximately 10

- the potential limit of the reaction due to the electrode system (e.g. for Ag/calomel it is between -50 and +50 mV) is adapted

- a dosing pump is connected with the Dulcotrol-Detox, which controls the addition of hydrogen peroxide

- the reaction will now be performed automatically by pressing the start key, passing the detoxification phase and will end at the potential adjusted before (plateau of the potential)

- in short time intervals the final potential is checked, and if there is a change, which sometimes occurs through dissociation of cyanide complexes, the device starts again into the detoxification phase until a stable potential can be reached.

If the Dulcotrol-Detox is connected with a recorder, the potential as well as the temperature can be drawn.

TABLE II: COST COMPARISON FOR CYANIDE DETOXIFICATION

| METHOD (REACTION PRODUCTS) | CONSUMPTION FOR 1 kg CN^- | | COSTS |
	CHEMICALS	ENERGY	DM/kg CN^-
ClO^- (NH_4^+; CO_2)	25 kg (13,5 wt%)	–	6,30 *
ClO^- (N_2; CO_2)	65 kg (13,5 wt%)	–	16,30 *
O_3 (N_2; CO_2)	3,6 kg	27–30 kWh **	14,– ***
electrolysis (N_2; CO_2)	–	40–50 kWh **	7,– ****
H_2O_2 (NH_4^+; CO_2)	5–6 kg (35 wt%)	–	9,60
Curox® (N_2; CO_2)	37 kg	–	200,–

*   decomposition during storage not calculated
**   energy costs only
*** including investment, o/m–costs, service of capital
**** theoretical value, highly dependent on current yield

FINAL DETOXIFICATION WITH CUROX^R

In our example (TABLE I) the final detoxification is performed by adding CUROX^R. This chemical compound, too, does not contain chloride and therefore does not chlorinate. However, as a reaction product sulfate is formed. The reaction [6] of the triple salt $2KHSO_5 . K-HSO_4 . K_2SO_4$ is

$$CN^- + HSO_5^- \longrightarrow CNO^- + HSO_4^- \tag{15}$$

$$2CNO^- + 3HSO_5^- + 2H^+ \longrightarrow 2CO_2 + N_2 + 3HSO_4^- + H_2O. \tag{16}$$

The application of CUROX^R is recommended for the treatment of low cyanide concentrations and if cyanide complexes (e.g. nickel complexes) are present, which cannot be oxidized completely with hydrogen peroxide. The reason for this limited application is an economical one due to the relatively high price of CUROX^R.

COST COMPARISON

According to the chemical reactions (1) to (16) chemical and energy costs are listed in TABLE II. The costs are calculated in DM/kg CN$^-$ under German circumstances, i.e. German prices. With the exception of CUROX^R the costs for oxidizing 1 kg CN$^-$ are in the same range and therefore comparable. The investment costs, however, for ozone or electrical devices are much higher than for very simple equipment (pumps, stirrer) needed for the application of hypochlorite and hydrogen peroxide. Taking into account that the stability during storage is much higher for hydrogen peroxide than for hypochlorite, the application of hydrogen peroxide may be a more cost-effective process for cyanide detoxification.

AOX

People in Germany are concerned about this parameter, because there is a suspicion that chlorinated organic compounds are toxic, carcinogenic and mutagenic. As a matter of fact, AOX forming compounds should be avoided or eliminated, because more and more surface water must be used for human purposes (drinking water, cooking, cleaning etc.). For this reason, local authorities in some parts of the country do not allow the use of chlorine or chlorine compounds for waste water treatment like cyanide detoxification. This was the reason for our investigations whether AOX is reduced or formed (de novo-AOX) during oxidation of cyanide in waste water with different processes. In TABLE III a number of data worked out in our laboratory using different samples of waste water are summarized. Obviously the AOX concentrations in the treated samples did not (except for a slight increase in sample 1 and 5 using H_2O_2/CUROX) increase after using hydrogen peroxide without or with CUROX^R, being in contrast with samples treated with hypochlorite. In some samples even a reduction of AOX could be observed. As a conclusion, de novo-AOX-formation is more probable using hypochlorite than it is using the active oxygen compounds H_2O_2 and CUROX^R.

TABLE III: AOX REDUCTION/FORMATION AFTER CYANIDE
DETOXIFICATION WITH DIFFERENT METHODS

SAMPLE	1	2	3	4	5
UNTREATED					
CN (mg/l)	1.270	9.000	780	170	1.850
AOX (μg/l)	55	150	320	210	525
TREATED					
CN (mg/l)	<0,1	<0,1	n.d.	n.d.	<0,1
AOX (μg/l)					
H_2O_2*	55	130	140		110
H_2O_2/Curox*	60	130	90	190	630
Curox				195	
ClO⁻	295	840	530	390	45.000(!)
ClO⁻/H_2O_2			500		

*   sample 3 and 5: addition of activator

SUMMARY

The cyanide detoxification with active oxygen compounds, i.e. hydrogen peroxide and/or
CUROX[R] is

-        technically feasible
-        easy to perform
-        controllable by the electrochemical potential
-        cost effective and
-        alleviates AOX formation.

ACKNOWLEDGEMENT

The author would like to thank all colleagues of INTEROX for useful discussions and sugges-
tions as well as the laboratory staff and the secretary for the careful and extensive work.

REFERENCES

[1]   Oehme, F. et al., Galvanotechnik + Oberflächenschutz, 7. Jhrg., Heft 3 (1966), p. 74-84

[2]   Bird, A.J., Birmingham University, Dept. Chem. Eng., Vol. 26, Nr. 1 (1975), p. 12-21

[3]   Roos, H. and Schmidt, M., Chem.Ing.Tech. 49, Nr. 5 (1977), p. 369-374

[4]   Fabjan, Ch. and Davies, R., Wasser, Luft und Betrieb, 20, Nr. 4 (1976), p. 175-178

[5]   Pini, G., Chemie-Technik, 4.Jhrg., Nr. 7 (1975), p. 257-260

[6]   Anon., Epuration des eaux résiduaires cyanurées par les percomposés, L'Air Liquide, Departement Chimique (1971)

Advanced Chemical Oxidation of Contaminated Water Using the perox-pure™ Oxidation System

ABSTRACT

The use of chemical oxidation processes to achieve on-site destruction or detoxification of contaminated waters and wastes will increase dramatically during the next decade. Advanced oxidation processes will compete with so called "transfer technologies," such as air stripping and activated carbon adsorption.

Development of the **perox-pure™** oxidation system began in the late 1970's, and there are now about forty (40) full-scale treatment systems in operation. Data and case histories from some of these treatment systems are presented, along with a discussion of the advantages of the **perox-pure™**[1] design. In addition to economic and performance data, recent process improvements that provide operating cost savings of up to 50% are highlighted. This new technology is demonstrated to improve the oxidation rate of difficult-to-oxidize chlorinated alkanes by up to 320%.

INTRODUCTION

The trend for treatment of toxic and hazardous organic wastes is toward total on-site destruction. During the 1980's new technology was developed to enable industry to meet the more stringent treatment standards for these toxic chemicals. Advanced chemical oxidation processes have led the way and are rapidly becoming the technology of choice for many applications. The decade of the nineties will most certainly see these processes become an established treatment technique as more full-scale systems become operational. This paper will discuss one such process - the UV/peroxidation Chemical Oxidation System. The focus will be on the development of the process and the experience gained in getting nearly forty (40) full-scale systems operational. Also included will be a discussion of process improvements that have led to significantly reduced costs and improved performance.

---

[1] **perox-pure™** - A UV/H_2O_2 chemical oxidation process using high intensity UV lamps.

Emery M. Froelich, Technical Director, Peroxidation Systems, Inc., 5151 E. Broadway, Suite 600, Tucson, Arizona, U.S.A.

UV/PEROXIDATION PROCESS

In this process, UV light converts the hydrogen peroxide (H_2O_2) in solution to hydroxyl radicals (HO•) and "activates" many of the organic molecules to make them easier to oxidize. The photolysis reaction which forms HO• can be shown as follows:

$$H_2O_2 \xrightarrow{\text{UV}} 2\ HO•$$

The activation of the organic molecules can range from direct oxidation by absorption and disassociation to the formation of organic radicals or other reactive intermediates. With enough time and reactants, organic compounds can be completely destroyed to CO_2, H_2O and, if present, the appropriate inorganic salt.

Most early UV oxidation processes used low pressure mercury vapor lamps combined with ozone (O_3). The UV/H_2O_2 process discussed in this paper utilizes a proprietary high intensity UV lamp combined with H_2O_2 (UV peroxidation). During the development of this process, the type of UV lamp and oxidant were evaluated to determine the optimum system. Two different types of UV lamps were available and are compared in Table 1. The "low intensity" (LO) lamp is the standard one used in disinfection systems with essentially all of the UV output at 254 nm. While the efficiency of conversion from electrical power to UV is relatively high, the maximum intensity is only about 1 watt per inch of lamp length and the efficiency drops significantly if the lamp is operated outside a narrow temperature range. The typical "high intensity" (HI) lamp has a somewhat lower conversion efficiency, around 15%, but proprietary modifications can boost this to nearly the same efficiency as the LO lamps. Benefits of the HI lamp include a broad spectrum UV output, relative insensitivity to temperature and maximum output of 200-600 watt/inch of lamp length.

TABLE 1

COMPARISON OF UV LAMPS

	Low Intensity	High Intensity
Output Efficiency	30-40%	15-35%
Spectrum	254 nm	180-400 nm
Temperature Effects	Sensitive	Not Sensitive
Output Energy	1 Watt/In	200-600 Watts/In
Lamp Cost/KW Output	$2,000-3,300	$90-220

The benefits of the HI lamp include lower capital cost per installed KW and lower maintenance requirements. In addition, the broad band spectrum output can "activate" or oxidize a wide range of organic compounds.

A similar evaluation was made of the type of chemical oxidant to use. Both ozone and hydrogen peroxide (H_2O_2) contain no halogens or other toxic by-products and can be decomposed to water and oxygen. Costs per pound of ozone and H_2O_2 are also very similar (Table 2). Ozone, however is not very soluble in water and must be generated on-site at a relatively high equipment and operating cost. Hydrogen peroxide is completely miscible with water, is commercially available in solutions of 30-50% by weight and requires only a small investment in storage and feed equipment.

TABLE 2

COMPARISON OF OXIDANTS

	Ozone	H_2O_2
Water Solubility	Very Low	Miscible
Cost/lb.	$0.6-1.10	$0.75-1.50
Equipment Cost (100 lb/day)	>$100,000	<$10,000

The basic design of the UV peroxidation system, proprietary HI lamps combined with H_2O_2, has the following advantages for on-site destruction of toxic organic contaminants.

- Low Capital Cost - Utilization of HI lamps and H_2O_2 reduces the capital cost required to treat a given waste stream - minimal equipment cost for oxidant and much lower cost per unit of UV power needed.

- Compact Design - A simple skid mounted unit with dimensions of 2' x 8' utilizing six (6) lamps can treat many contaminated groundwater flows up to 400 gallons/min. A comparable UV/H_2O_2 system using LO lamps would require over 1000 lamps!

- Rapid Reaction - Reaction rate increases with increased UV intensity (Figure 1). By using HI lamps even difficult to oxidize compounds can usually be oxidized in only a few minutes.

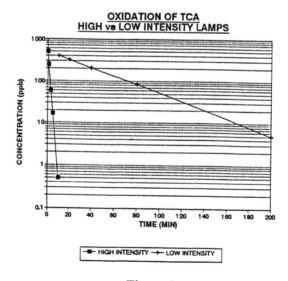

Figure 1

- No Gas Emissions - Ozone, being a toxic gas, requires a special gas phase decomposition system and sometimes an air discharge permit. By utilizing H_2O_2 and no sparging, a UV peroxidation system avoids any possible air emissions.

- Wide Range of Applications - The combination of HI lamps and H_2O_2 allows for rapid oxidation of contaminants where untreated concentrations range from the parts per billion range up to the low percent organic level.

One additional problem that is avoided by using H_2O_2 is illustrated in Table 3. Difficult to oxidize volatile organics such as 1,1,1-trichloroethane (TCA) take a very long time to oxidize in a UV/ozone system because of the LO lamps and the limited solubility of ozone. These same compounds, however, are easily stripped from the water. Table 3 shows some comparison data from treatment of a contaminated groundwater using UV/ozone and the UV peroxidation process. Most of the TCA is stripped from the water by the ozone before it can be oxidized. The UV peroxidation process, however, is able to completely oxidize TCA without stripping it into the air.

TABLE 3

REMOVAL OF 1,1,1-TCA
UV/OZONE VS UV PEROXIDATION

	UV/O_3[7]	UV peroxidation
Reaction Time 90% Removal	>40 Min.	3-4 Min.
% Oxidized	0-30	90
% Stripped	60-90	0

DESIGN AND OPERATION

While advanced chemical oxidation processes are becoming widely accepted for removal of a broad range of organic contaminants, they are not like the proverbial "black box" that you simply connect to any contaminated water and say the magic word. A successful treatment system that achieves consistent performance must take into account any and all operational variables that can affect the system design.

CONTAMINANT CONCENTRATIONS

Influent and effluent contaminant concentrations are often poorly quantified. When faced with this unknown variability it is standard practice to design for "worst case" conditions. This doesn't mean, however, that you should automatically size the advanced oxidation system to handle the maximum influent concentration conceivable and the minimum effluent criteria that could take effect 5-10 years down the road. Equalization, staged start-up of a well field, and/or modular treatment units that can be easily upgraded should be considered to optimize the cost-effectiveness of the treatment system. The UV peroxidation process can handle a wide range of organic compounds and concentrations, but the reaction does follow a pseudo first order rate. This means that the UV energy required to remove 90% of a contaminant is the same regardless of the starting concentration.

INORGANIC CHEMICALS

Most soluble inorganic chemicals do not have a significant impact on the design and operation of the UV peroxidation process. High brine wastes may slow the oxidation rate somewhat, but that can be evaluated during a bench-scale test. One area that is often overlooked is water stability. It is well known that some soluble inorganic chemicals such as calcium and iron will precipitate out of solution with only small changes in temperature, pH, etc. of some waters. This problem, if ignored, has resulted in poor performance in air strippers and steam strippers and plugged and/or solidified carbon columns and sand filters. The effect of this phenomenon on UV oxidation systems is to coat the quartz tubes and reduce performance through blocking the UV light penetration. Scraping, mechanical cleaning, acid washing or other chemical treatment can be effective in removing the scaling caused by an unstable water, but the best method is to prevent the problem from occurring by stabilizing the water before treatment.

OXIDANT AND CATALYST DOSE

It may be tempting to view the addition of H_2O_2 like taking a medication. If you add some and the results are good, perhaps you should add a lot more to get better results. Not only is this expensive, but can sometimes give the exact opposite results. The oxidation of a number of organic contaminants require "activation" by direct UV light absorption. Excess H_2O_2 can absorb the energy inhibiting this UV activation and resulting in poorer performance.

Catalyst addition, if beneficial, must also be carefully controlled. Our experience indicates that only a small percentage of contaminated waters can benefit from catalyst addition. Those that do show a benefit require a carefully controlled addition to maximize the benefit.

PH

Some organic contaminants oxidize faster at a specific pH. Some catalyst systems work better below a certain pH. In addition to these parameters, which can be optimized during bench-scale testing, pH adjustment can also improve the stability of some waters and reduce interferences caused by some compounds. For these reasons it is important to control pH in many applications. This may add to the operational complexity and equipment cost, but it is not unusual to reduce operating costs by 40-65%.

PRETREATMENT

A large fraction of the operating problems and inconsistent results obtained in water and wastewater treatment process can be traced to inadequate or overlooked pretreatment requirements. Our experience with the UV peroxidation process supports this statement. While UV oxidation can be carried out in the presence of some amount of suspended solids, color and unstable water, the efficiency and costs can be significantly improved by a correctly designed pretreatment system

- <u>Solids</u> - Filtration and/or clarification are established processes that can be easily incorporated to remove suspended solids and turbidity.

108

- Water Stability - Many groundwaters and some wastewaters can cause scaling or coating of wetted surfaces with iron and calcium deposits. Small changes in oxygen levels, pressure, pH or even just the presence of a large surface area can result in scaling. In some cases the process of destroying the organic compounds may result in precipitation of chelated metals. A well designed pretreatment system incorporating pH adjustment, precipitation or stabilizing chemical addition will prevent scaling and give more consistent results than post-scaling cleaning systems.

- Other Interferences - Some inorganic and organic species such as color can be controlled by traditional pretreatment systems or pH adjustment. A more complex problem is caused by high levels of "non target" organic compounds. These compounds are usually considered non toxic or easily biodegradable. In some cases the compounds are considered contaminants but can be recovered if the more toxic low concentration contaminant can be removed. Pretreatment in this case is more appropriately considered combined treatment. Treatment processes such as Fenton's oxidation, steam stripping, ultrafiltration or gas phase catalytic oxidation should be considered in combination with UV peroxidation.

An example of the benefits of a well designed pretreatment system is shown in Table 4. This data is from a full-scale UV peroxidation system treating contaminated groundwater. This water contained up to 20 mg/l of iron and high calcium hardness. Before pretreatment the iron and calcium would precipitate during the oxidation process and coat the quartz tubes. A pre-oxidation/filtration system was added and the water stabilized by adjusting pH. Once this pretreatment system was operating properly, the results have consistently met the effluent criteria.

TABLE 4

FULL-SCALE TREATMENT RESULTS
WITH UV PEROXIDATION

| | Influent (μg/l) | Effluent (μg/l) | |
		No Pretreatment	w/Pretreatment
Benzene	7,600	1,200	ND
Toluene	24,000	1,300	ND
Cl-Benzene	8,800	1,100	ND
Eth-Benzene	3,300	260	ND
Xylenes	46,000	1,900	ND

ND = Non Detectable.

PROCESS IMPROVEMENTS

The UV peroxidation process has proved to be a cost-effective treatment system for many groundwater and wastewater streams. On-going development work has focused on improving the process design with particular emphasis on the "difficult to oxidize" compounds such as 1,1,1-TCA, 1,1-DCA, MeCl and $CHCL_3$. Recent improvements in reactor design and UV output efficiency have reduced operating costs for some compounds by 50%.

The improvement in oxidation rate of some selected compounds is shown on bench-scale equipment in Figures 2-4. This data was generated using spiked tap water solutions containing mixtures of 3-5 volatile organic chemicals at an initial total organic concentration of 5-15 mg/l. In all cases the oxidation rate increased by 2-3 times with the new B design. Similar testing done on clients groundwater and wastewater samples show the same range of oxidation rate improvement with these "difficult to oxidize" chemicals.

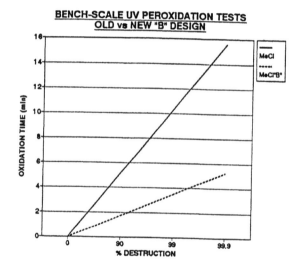

Figure 2

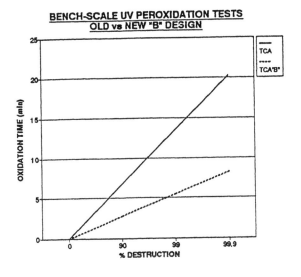

Figure 3

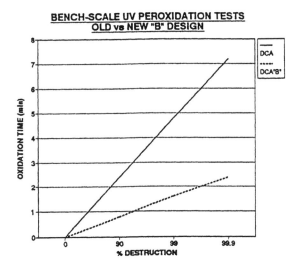

Figure 4

The impact of this new technology is shown in Table 5. A UV peroxidation treatment facility in the southwest was converted to this new "B" design. Both capital and operating cost reductions were achieved due to the improved oxidation efficiency.

TABLE 5

FULL-SCALE UV PEROXIDATION TREATMENT
SITE 1

	"A" Design	"B" Design
Flow Rate	150 gpm	150 gpm
Contaminants	TCE, DCE	TCE, DCE
Capital Costs	$150,000	$100,000
Operating Costs/1000 gal	$0.97	$0.25

CASE HISTORIES

To further illustrate the main points presented in this paper, Tables 6 and 7 present data from two examples of full-scale operating UV peroxidation treatment systems. The system operating in the northeast at Site 2 (Table 6) treats a contaminated groundwater that is treated and reused at the site. While there are about a dozen contaminants in the water, only methylene chloride (MeCl) and TCA are shown as they are the most difficult to remove. Some pretreatment for solids removal was incorporated in the original system design. By improving the pretreatment system to eliminate some further interferences and stabilize the water, more consistent and better removal of all contaminants was achieved.

TABLE 6

FULL-SCALE UV PEROXIDATION TREATMENT
SITE 2

	Influent ($\mu g/l$)	Effluent ($\mu g/l$)	
		Before Pretreatment	After Pretreatment
MeCl	903	89	11
1,1,1-TCA	60	20	6

The second example, shown in Table 7, is presented to illustrate the ability of the UV peroxidation process to handle interfering compounds such as iron and copper. By paying careful attention to the water chemistry and pretreatment needs, good oxidation was achieved without precipitation of iron or copper.

TABLE 7

FULL-SCALE UV PEROXIDATION TREATMENT
SITE 3

			Influent	Effluent
A.		Hydrazine	150 mg/l	<0.1 mg/l
		EDTA	450 mg/l	<10 mg/l
		Iron	15 mg/l	15 mg/l
B.		EDA	1400 mg/l	<100 mg/l
		EDTA	50 mg/l	<10 mg/l
		Copper	5 mg/l	5 mg/l

Because of the high levels of organic chealating agents, the iron and copper can't be removed before UV peroxidation treatment. Keeping the pH below 4 during treatment prevents the metals from precipitating during oxidation. After the organics are removed, raising the pH followed by clarification removes the metals.

SUMMARY

The decade of the 1990's will see advanced chemical oxidation processes become the technology of choice for on-site destruction of toxic organic contaminants. Installation and operation of nearly forty (40) full-scale UV peroxidation systems demonstrate the cost-effectiveness of this process. This experience clearly shows the importance of evaluating and controlling critical operational parameters, including pretreatment, to achieve consistent performance.

REFERENCES

1. Bernardin, F.E., Jr., 1990. "UV/Peroxidation Destroys Organics in Groundwater", Presented at the 83rd Annual Meeting of the Air and Waste Management Association, Pittsburgh, PA.

2. Cheuvront, D.A. and Swett, G.H., 1988. "Innovative Groundwater Treatment Technologies with Zero Air Emissions", Hazmacon 88, Anaheim, CA.

3. Froelich, E.M., 1990. "The **perox-pure**™ Oxidation System - A Comparative Summary," The Symposium on Advanced Oxidation Processes, Toronto, Canada.

4. Glaze, William H., Joon-WUN-Kang, Chapin, Douglas H., 1987 "The Chemistry of Water Treatment Processes Involving Ozone, Hydrogen Peroxide and Ultraviolet Radiation", Ozone Science & Engineering, Volume 9 pp. 335-352.

5. Hager, D.G. and Smith, C.E., 1986. "The Destruction of Organic Contaminants in Water by Chemical Oxidation", Haztech International, Denver, CO.

6. Hager, D.G., Loven, C.G. and Giggy, C.L., 1988. "On-site Chemical Oxidation of Organic Contaminants in Groundwater Using UV Catalyzed Hydrogen Peroxide," AWWA Annual Conference and Exposition, Orlando, Florida.

7. Lewis, Norma, Topudurti, Kirankumar and Foster, Robert, 1990. "A Field Evaluation of the UV/Oxidation Technology to Treat Contaminated Groundwater," HMC, pp. 42-44.

Design of Completely Mixed Ozonation Reactors

ABSTRACT

Ozonation is a technically viable wastewater treatment alternative. The process is a simultaneous mass transfer - chemical reaction process. Because of the inability to adequately describe the kinetic mechanism when other pollutants exist and daughter products are formed, feasibility data are required to determine global rate constants for design. An example of a design problem is given.

INTRODUCTION

Ozone, O_3, is an extremely strong oxidizer. Consequently it has a wide variety of applications in the chemical processing sectors of industry. Ozonation has also been shown to be a technically feasible alternative to treat a wide range of inorganic, organic, and biological species in water.

Ozone may be practical for treating isolated streams as a pretreatment to remove refractory species or to pretreat refractory organics as a polishing step or to partially oxidize refractory organics, increasing the biodegradability without significant reduction of total organic carbon. Experience has been obtained using ozone with ultraviolet light (UV). The UV enhances the kinetics of decomposition for a number of organic compounds. It has also been investigated for use as a pretreatment for carbon adsorption. Extensive work has been performed on a wide range of wastewaters of industrial interest. A

John A. Roth, Professor, Chemical and Environmental Engineering, Vanderbilt University, Nashville, Tennessee, U.S.A.

comprehensive review of the state of the art is given by Rice and Browning (1980), who classify the reported work in twenty-one industrial categories. Their review is summarized in Table 1.

Systems for ozonation of wastewater have four major components (Rice and Browning, 1980):

1. electrical power generation
2. air or oxygen feed preparation
3. ozone generation
4. ozone-wastewater contacting

The design parameters of the ozone generation equipment is established by the equipment manufacturers. An extensive discussion of the variables involved is given by Rosen (1972, 1973) and Masschelein (1982).Evolving technology has increased the efficiency of ozone generation which has increased today to above 24 kwhrs/kg of ozone generated. This paper considers the process design for the removal of pollutants.

THEORY

When a gas stream containing pure ozone is contacted with a wastewater containing a pollutant, the phenomena which occur are:

o mass transfer of the ozone from the gas phase to the liquid
o diffusion in the aqueous phase to a pollutant species
o self-decomposition of the ozone (under certain conditions of pH and temperature)
o reaction of the ozone (or free radical oxidizer species formed) with the pollutant
o further oxidation of the daughter products of the original pollutants down the oxidation chain
o simultaneous removal of volatile organics by gas stripping

The theory for mass transfer with simultaneous reaction has been discussed in detail (Asterita, 1967; Carberry, 1976; Charpentier, 1981). For relatively simple reaction mechanisms and using the two film theory, rate expressions can be developed for the following regimes: very slow reaction in bulk liquid; slow reaction in bulk liquid; moderately fast reaction; and fast reaction in the diffusion film. The fast reaction in the diffusion film may be pseudo-first order, in which case the rate of absorption is an apparent half order with respect to the pollutant concentration. Sotelo et. al. (1989) showed that azo dye ozonation followed this a parallel reaction mechanism and that the reaction was pseudo-first order for direct yellow 27 and acid black dyes, but not for direct blue dye under the conditions at which the reactions were carried out. He used an Arrhenius temperature relationship and an empirical power function to fit the effect of [OH⁻]. The ozone diffusivity can be obtained from the expression (Matrozov et. al., 1976).

$$D_{O3}\mu/T = 4.27 \text{ x } 10^{-15} \tag{1}$$

where

μ = viscosity, kg m^{-1} s^{-1}

T = Temperature °K

The equilibrium Henry's law constants can be determined experimentally using the method of Roth and Sullivan (1981) or estimated from their correlation:

$$H = 3.84 \text{ x } 10^{-7} \text{ [OH}^-\text{]}^{0.035} \exp (-2428/T) \tag{2}$$

where

H = Henry's law constant, atm/mole fraction

[OH⁻] = hydroxide concentration (g-mol/L)

Mass transfer coefficients may be estimated from correlations of systems similar to the contactor to be used or determined experimentally.

ENGINEERING PRINCIPLES

When a reaction is carried out in a semibatch process, the reaction proceeds rapidly. As more and more of the pollutant is oxidized towards completion, the reaction becomes much slower. During the course of a semibatch reaction, it is possible that a reaction may move through two or more of the mass transfer-reaction controlling regimes. It is also significant that other ozone reactive species and daughter products may compete with the primary pollutant(s) for the ozone. Furthermore, the ozone will self decompose at higher pH's and temperatures. Accordingly, the process becomes much more difficult to describe mechanistically. For continuous flow reactors, a steady state material balance on the component being treated and assuming power kinetics of order n, the global rate constant is given by:

$$k_{(f)n} = \frac{Q(C_o-C)}{VC^n} \tag{3}$$

where

$k_{(f)n}$ = global rate constant

116

n = apparent reaction order

C = outlet concentration of the pollutant

C_o = inlet concentration of the pollutant

Q = liquid volumetric flow rate

V = reactor volume

For a semi-batch reactor, the kinetic order is determined using a least squares regression fit of the appropriate integrated form of the material balance expression:

$$-\frac{dC}{dt} = k_{(s)n}\ C^n \tag{4}$$

where

C = pollutant concentration in the reactor

t = time

The global rate constant, $k(s)_n$, contains the effect of the intrinsic kinetics and the mass transfer, and may reflect more than one regime mass transfer-chemical reaction regime. The global rate constant, $K(f)_n$, in a completely mixed vessel, reflects only one reaction regime. A schematic of a bench scale ozonation system is shown in Fig. 1. Whitlow and Roth (1988) reviewed the rate data of approximately twenty investigators from the literature who had reported sufficient data to quantify. The overall order and a global rate constant for the disappearance of the pollutants was determined. They found that phenol was first order and that cyanide destruction was zero order. They further fit the global rate constants to an equation of the form:

$$K_n = \alpha\ (G/V)^x\ ([OH^-])^y\ (C_\phi OH)^z \tag{5}$$

where

G/V = ozone application rate

$[OH^-]$ = hydroxyl ion concentration

C_o = initial concentration of the pollutant

α, x, y, and z = empirically determined constants

Figures 2 and 3 show the results of these correlations of the data of Li (1979), Gurol (1980), Moench (1980), and Eisenhauer (1968). These results show consistency between these investigators. These results were all obtained from semibatch experiments. Only limited data were found for completely mixed flow reactors. Correlations from these data followed the same empirical form as the semibatch reactors.

DESIGN PROBLEM

A dye-stuffs plant outputs a liquid waste stream of 200,000 gal/day, which contains 225 mg/l of Solophenyl blue dye. An ozone dye-wastewater treatment unit is designed to bring the dye concentration in this stream within the effluent standard of 10 mg/l.

CHOICE OF REACTOR

In industrial situations, continuous processes are generally preferred over batch ones. This reduces the size need for the storage/overflow facility. Also, more continuous processes are more easily tuned to a higher degree of control. For this reason, a continuous stirred tank reactor, CSTR, is selected for the dye removal process.

A single glass-lined, or stainless-steel reactor is chosen. The reactor liquid height to diameter ratio is designed to be 1.5, since the design equation is based on experimental runs with this geometry.

OZONE PRODUCTION

The ozone is produced by passing a gas stream, containing oxygen, between the plates of a capacitor, in a commercial ozone generator. The feed gas may be air, pure oxygen, or an ozone-oxygen mixture. The latter is possible when a recycle stream, containing some unused ozone, is mixed with pure oxygen and passed through the generator again. The feed gas is necessarily bone-dry, so an adsorptive dryer (dessicant) is placed upstream from the ozone generator.

According to Evans (1972), "(The) recycling of the oxygen (is) an economic necessity." However, during each pass through the reactor, the gas stream picks up nitrogen which is dissolved in the liquid. A buildup of nitrogen then occurs in the loop, and eventually has adverse effects on the ozone generator. A continuous purge must be provided to control the nitrogen concentration. The production cost of ozone is a function of the ozone concentration, G.

TEMPERATURE AND pH EFFECT

The design equation is applicable for a temperature of 25°C, and this temperature will be used to design the unit. The effect of temperature is complicated: Qualitatively, the mass transfer increases slightly with temperature and the solubility of ozone decreases with temperature. If the reaction is slow and in the kinetic regime, the temperature effect will follow an Arrhenius form, and will significantly affect the reaction rate. This effect must be determined experimentally for a particular wastewater and can be included in the multilinear regression of a form of equation.

The rate constant is a weak function of the hydroxyl ion concentration for this waste. The waste stream will be neutralized (pH = 7), upstream from the reactor, to minimize control problems.

Reactor Volume and Ozone Application Concentration

The design equation is used to relate the reactor liquid volume and the ozone application concentration. A material balance over a CSTR yields the following equation:

$$r_A = \frac{Q(DYE)_o - (DYE)}{V} \qquad (6)$$

The (G/V) term in equation (4) was modified for the experimental data obtained to develop a modified correlation based on the ozone dose rate. The global rate constant correlation then becomes:

$$r_A = .000493\, w^{.635}\, [OH^-]^{-.019}\, (DYE)^{.210}\, (DYE)^{2/3} \qquad (7)$$

Combining equations (6) and (7) yields:

$$V = \frac{Q[(DYE)_o - (DYE)]}{.000493\, w^{.635}\, [OH^-]^{-.019}\, (DYE)_o^{.210}\, (DYE)^{2/3}} \tag{8}$$

where

    w   = ozone appl. conc., mg/L

    V   = reactor, volume, L

    Q   = liquid vol. flow rate, L/sec

    (OH^-) = hydroxyl ion conc., gm mol/L

    (DYE) = dye concentration, mg/L

    $(DYE)_0$ = initial dye conc., mg/L

The index of determination of this correlation, based on the experimental data is 0.88.

Four of the variables in equation (8) are already specified.

    Q   = 200,000 gal/day - 8.82 L/sec

    $(DYE)_o$= 225 mg/L

    (DYE) = 10 mg/L

    (OH^-) = 1 E-7 mol/L (pH = 7)

Substituting these numbers into equation (8) yields:

$$V = \frac{196,000}{w^{0.635}} \tag{9}$$

Equation (9) illustrates the trade-off which exists between reactor volume and ozone application concentration. The costs of the ozonation reactor are directly related to the reactor volume.

A detailed optimization is not performed since sufficient economic data are not directly available. Therefore, the design is carried out for an ozone application concentration of 2 percent, by weight, at standard temperature and pressure [as recommended by U. S. Ozonair

Corp., (1978)]. This corresponds to 24 mg/L ozone concentration, again at standard temperature and pressure. The liquid reactor volume necessary is calculated, from equation (9), to be 26,100 liters. For an error margin, 10 percent is added to this value; thus the liquid volume is 28,700 liters, or 1,000 ft^3. The total reactor volume includes 33 percent freeboard. Therefore, the total volume is 10,000 gallons, or 1,340 ft^3. The dimensions of the reactor are calculated as follows:

For the Liquid

$H/D = 1.5$

$\pi D^2 H/4 = 1,000$ ft^3

$D = 9.5$ ft. = reactor diameter

For the entire Reactor

$\pi D^2 H'/4 = 1,340$ ft^2

$H' =$ reactor height = 18.9 ft.

In summary, the reactor has a total capacity of 10,000 gallons; it is 9 ft. 6 in. in diameter, and 19 ft. high. The operating liquid height inside the reactor is approximately 14 ft.

The final design parameter is the gas flow rate, w. The ozone application rate to liquid reactor volume ratio, wG/V, is fixed as a process parameter, since it was fixed in the experimental work.

$$wG/V = 60\,mg/L\cdot min = 2.1\,mg\cdot ft^3/L\cdot min \qquad (10)$$

Therefore,

$$G' = 2.1\,mg\cdot ft^3/L\cdot min\,\frac{(28700\,L)}{(24\,mg/L)} = 2500\,SCFM \qquad (11)$$

It is necessary to produce 2,500 SCFM of 2 percent, by weight, ozone in oxygen and to sparge it into the bottom of the reactor, through ceramic diffusers. A summary of the design criteria are given in Table 2.

Materials of Construction

Ozone is an extremely powerful oxidizer. Consequently, the proper selection of materials of construction is mandatory. Monroe and Key (1981) give some recommended materials of construction for ozonation systems. These are shown in Table 3.

REFERENCES

[1] Astarita, G., <u>Mass Transfer with Chemical Reaction</u>, Elsevier Publishing Company, New York, 1967.

[2] Carberry, James J., <u>Chemical and Catalytic Reaction Engineering</u>, McGraw-Hill Book Company, New York, 1976.

[3] Charpentier, Jean-Claude, "Gas-Liquid Absorptions and Reactions," in <u>Advances in Chemical Engineering</u>, Edited by T. B. Drew et. al., Academic Press, New York, 1981.

[4] Eisenhauer, H. R., "The Ozonation of Phenolic Wastes," J. Water Poll. Control Fed, 40:11, Part 1, 1887, 1968.

[5] Evans, F. L. III, editor, <u>Ozone for Water and Wastewater Treatment</u>, Ann Arbor Science Publishers, Inc., Ann Arbor, Michigan, 1972.

[6] Gurol, M. D., "Kinetic Behavior of Ozone in Aqueous Solution: Decomposition and Reaction with Phenol," Ph.D. Dissertation, University of North Carolina, 1980.

[7] Li, K. Y., "Kinetic and Mass Transfer Studies of Ozone-Phenol Reaction Systems," Ph.D. Dissertation, Mississippi State University, 1977.

[8] Masschelein, W. J., <u>Ozonation Manual for Water and Wastewater Treatment</u>, John Wiley & Sons, New York, 1982.

[9] Matrozov, V., S. Kachtunov, A. Stepanov, B Tregunov, "Experimental Determination of the Molecular Diffusion Coefficient of Ozone in Water," Zh. Prikl. Khim. 49:1070-1073, 1976.

[10] Moench, W. L., "The Kinetic Modeling of the Ozone and Phenol Reaction System," Master of Science Thesis, Vanderbilt University, Nashville, Tenn., 1980.

[11] Rice, R. G. and M. E. Browning, "Ozone for Industrial Water and Wastewater Treatment," EPA Report No. 600/2-80-060, April 1980, U.S. Environmental Protection Agency, Robert L. Kerr Environmental Research Laboratory, Ada, Oklahoma 74820.

[12] Rosen, H. M., in <u>Ozone for Water and Wastewater Treatment</u>, F. L. Evans III, editor, Ann Arbor Science Publishers, Inc., Ann Arbor, Michigan, 1972.

[13] Rosen, H. M., 1973, "Use of Ozone and Oxygen in Advanced Wastewater Treatment," J. Water Poll. Control Fed. 45(12): 2521-2536.

[14] Roth, J. A. and D. E. Sullivan, "Solubility of Ozone in Water", Ind. and Eng. Chem. Fund., 20:137-140, 1981.

[15] Sotelo, J. L., F. J. Beltran, J. Beltran-Heredia, and J. M. Encinar, 1989 "Azo Dye Ozonation Film Theory Utilization for Kinetic Studies," J. Int. Ozone Assoc., 391-410.

[16] U. S. Ozonair Corp. Bulletin, "Comparative Data Between Old and New Ozone Equipment and Technology," U.S. Ozonair Corporation, San Francisco, 1978.

[17] Whitlow, J. E. and J. A. Roth, "Hetrogeneous Ozonation Kinetics of Pollutants in Wastewater," Environmental Progress, 7:52-57, 1988.

TABLE I.	INDUSTRIAL CATEGORIES REPORTING THE USE OF OZONE[a]	
Category	**No. of Articles Found**	
	Actual Wastes[b]	Total
Aquaculture	19	39
Breweries (under Food & Kindred Products)	5	9
Biofouling Control	4	10
Cyanides & Cyanates	--	24
Electroplating	17	27
Food & Kindred Products (except Breweries)	5	9
Hospitals	16	30
Inorganics	3	5
Iron & Steel	10	14
Leather Tanneries	0	3
Mining	11	15
Organic Chemicals	25	60
Paint & Varnish	4	4
Petroleum Refineries	22	27
Pharmaceuticals	0	2
Phenols	--	29
Photoprocessing	8	12
Plastics & Resins	4	4
Pulp & Paper	37	60
Soaps & Detergents	7	12
Textiles	<u>30</u>	<u>35</u>
TOTALS	227	430

[a]after Rice and Browning (1980)
[b]also. actual process waters

```
            TABLE II.   DESIGN SUMMARY

Influent:              Flow rate:  200,000 gal/day
                       Solophenyl blue dye concentration:
                       225 mg/l

Effluent:              Flow rate:  200,000 gal/day
                       Solophenyl blue dye concentration:
                       10 mg/l

Gas Flow:              Flow rate:  2,500 SCFM
                       Ozone concentration:  25 mg/l, or
                       2% (wt.) at standard temperature
                       and pressure

CSTR Reactor:          Volume:  10,000 gals. or 12,340 ft³
                       Diameter, 9 ft., 6 in.
                       Total Height:  19 ft.
                       Liquid Height: 14 ft.
                       Glass-lined or stainless-steel
                       construction
                       ceramic diffusers

Temperature:           25°C

pH:                    7.0
```

TABLE III.   OZONE SYSTEMS MATERIALS OF CONSTRUCTION
(Monroe and Key, 1981)

Type 304 or 316 stainless steel	Gas piping from inlet filter, through air preparation unit, through ozone generator to contact tank diffusers; contact vessels; water gates and weirs, ozone sample lines
Aluminum	Gas piping from inlet filter through air preparation unit to ozone generator
Aluminum oxide	Diffusers
Unplasticized PVC	Ozone sample lines
Concrete	Contact and holding vessels, channels, etc.
Teflon	Ozone sample lines
Hypalon	Gaskets
Viton	Gaskets

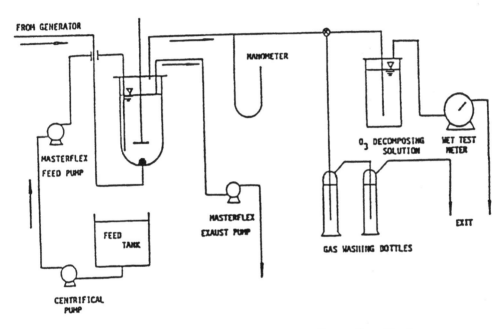

Figure 1. Schematic of Bench Scale Ozonation System

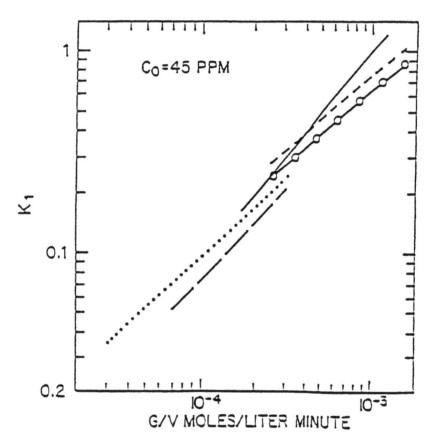

Figure 2. Comparison of the regression models for the semi-batch ozonation of phenol using first order kinetics at constant pH and initial concentration.

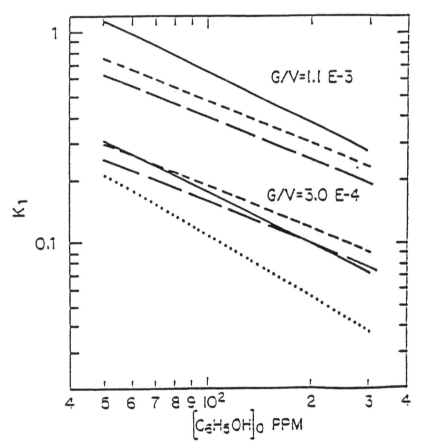

Figure 3. Comparison of the regression models for the semi-batch ozonation of phenol using first order kinetics at constant pH and ozone application rate.

Ozonation of Volatile and Semi-Volatile Compounds for Groundwater Remediation

ABSTRACT

Many of the U.S. EPA's list of target volatile and semi-volatile priority pollutants can be removed by chemical oxidants such as ozone (O_3) and hydrogen peroxide (H_2O_2). In the treatment reactor, the organic compounds are oxidized to various intermediates and end-products and/or volatilized to the gas phase during mixing or ozonation. The overall substrate removal process therefore, depends on the rate of volatilization as well as the chemical oxidation kinetics. This paper presents a methodology and develops a conceptual mathematical model for predicting the environmental fate of synthetic organics during treatment by chemical oxidation processes. It focuses on the use of ozone in bubble-type gas-liquid contactors since these units will exhibit significant removals by both oxidation and volatilization mechanisms. The general principals developed herein however, are applicable to other chemical oxidant systems if appropriate oxidation kinetic data are available. The proposed model and supporting analysis will facilitate identification and control of key process design variables to provide cost-effective treatment of wastestreams containing specific volatile and semi-volatile compounds.

BACKGROUND

It has long been recognized that volatile solutes are transferred to the atmosphere during mixing and aeration of wastewater. Historically, this removal pathway has been unregulated and frequently provided the primary means of treatment to satisfy regulatory discharge criteria for liquid wastestreams containing volatile and certain semi-volatile organic compounds. This treatment strategy is being eliminated however, by comprehensive regulation of all environmental media (air, liquid and solids). This now requires both off-gas and liquid phase management of treatment processes for these wastestreams. Although the emphasis for control of volatile compounds has been placed on listed priority substances many non-listed compounds must also be managed to satisfy "lumped" regulatory parameters for total volatile organic substances (TVOS) and for odor control. In order to effectively design treatment and control processes for these wastewaters a means for predicting the environmental fate of the target substrates is required.

Jack Musterman, Ph.D., P.E., Vice President and Director, Wastewater Management Division, Eckenfelder Inc., Nashville, Tennessee, U.S.A.

Numerous investigators[1,2,3] have developed mathematical models to determine the fate of organic priority pollutants during activated sludge treatment. The models typically describe the concurrent equilibrium relationships for substrate removal by bio-oxidation, floc sorption, and volatilization mechanisms. Many of the synthetic organic pollutants can also be effectively removed from wastewater by chemical oxidants such as hydrogen peroxide (H_2O_2) and ozone (O_3). In ozone reactors, organic compounds are chemically oxidized to various intermediates and end-products. They can also be transferred from the water to the gas phase during the gas-water contact period. The overall substrate removal process, therefore depends on the rate of volatilization as well as the chemical oxidation kinetics.

Until recently, chemical oxidation treatability tests have frequently sought to simulate a set of system-specific wastewater and process operating conditions. They typically measured total substrate or oxidant removal kinetics and did not differentiate between the kinetics of substrate oxidation and mass transfer of ozone/substrate in water. Studies of ozonation of synthetic organics by Gurol[4] and Hoigne and Bader[5,6] however, have developed fundamental kinetic data for ozone oxidation of synthetic organics in laboratory waters. These results can be combined with gas-liquid mass transfer relationships to estimate the environmental fate of synthetic organics during treatment by ozonation. This will facilitate identification and control of those process design variables to provide cost-effective treatment of wastestreams containing specific organic compounds.

OBJECTIVE

The purpose of this paper is to present a methodology and develop a conceptual mathematical model for estimating the removal of synthetic organic compounds from liquid wastestreams treated by chemical oxidation processes. It focuses on the use of ozone in bubble-type contactors since these units will likely exhibit significant removals by both oxidation and volatilization mechanisms. Although this paper has been developed using kinetic data for ozonation of synthetic organics, the general principals developed herein are applicable to other chemical oxidant systems if appropriate oxidation kinetic data are available.

DEVELOPMENT OF GENERAL VOLATILIZATION-OXIDATION MODEL

The generalized volatilization-oxidation model is based on materials balances for substrate and oxidant around the reactor system shown in Figure 1. The reactor is assumed completely mixed, covered and at steady state conditions with respect to both the liquid and gas phases. The input liquid stream contains the substrate at concentration $[S_i]_o$. It may or may not contain the oxidant depending on the chemical oxidation system selected. In a chlorine or hydrogen peroxide system the oxidant would be in the liquid stream. In an ozonation-bubble contactor system however, the ozone would be added as a separate gas stream.

The rate of volatilization of substrate, can be determined by mass balance for the gas phase in Figure 1 assuming that it is the only reaction contributing to the removal of substrate.

$$(r_i)_v = Q_L ([S_i]_o - [S_i]_e) = Q_g [S_i]_g \qquad (1)$$

Where

$(r_i)_v$ = Rate of removal of substrate i by volatilization, m/s

Q_L = Liquid influent and effluent flow rate, L/s

Q_g = Gas inflow and effluent flow rate, L/s

$[S_i]_o$ = Influent liquid phase concentration of substrate i, m/L

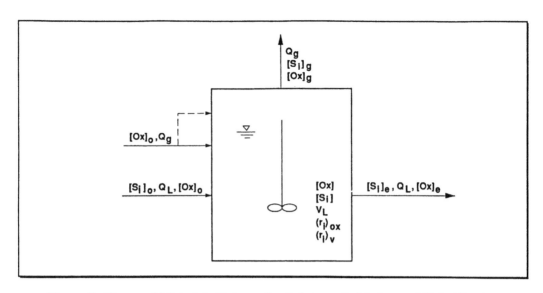

Figure 1. Flow and Materials Balance for Substrate Oxidation and Volatilization

$[S_i]_e$ = Effluent (and reactor) liquid phase concentration of substrate i, m/L

$[S_i]_g$ = Effluent and reactor headspace gas concentration of substrate i, m/L

The concentration of substrate in the gas phase depends on its volatility as expressed by Henry's Law and its mass transfer rate between the liquid and gas phases. The rate of volatilization can then be expressed in general terms as:

$$(r_i)_v = Q_g[S_i]_g = Q_g[S_i]_e H_i [f(K_L a)_i] \qquad (2)$$

Where

H_i = Henry's constant for substrate i, dimensionless
   = H_m/RT

$f(K_L a)_i$ = Functional relationship for mass transfer rate of substrate i between liquid and gas phases, 1/s

Combining Equations (1) and (2) provides an expression for the fraction of substrate remaining in solution if volatilization is the only removal mechanism.

$$([S_i]_e/[S_i]_o)_v = \frac{1}{[(Q_g/Q_L)\,(H_i)\,[f(K_L a)_i] + 1]} \qquad (3)$$

The rate of chemical oxidation of substrate is typically a function of both the substrate concentration and the oxidant concentration and can be expressed as:

$$(r_i)_{ox} = (k_i)_T [S_i]_e^m [Ox]^n \qquad (4)$$

Where

$(r_i)_{ox}$ = Rate of removal of substrate i by chemical oxidation, m/s

$[S_i]_e$ = Liquid phase concentration of substrate i in reactor, m/L

$[Ox]$ = Liquid phase concentration of oxidant in reactor, m/L

n,m = Empirically determined reaction coefficients

$(k_i)_T$ = Reaction rate coefficient

Term "$(k_i)_T$" represents a lumped rate coefficient that includes both direct oxidation and indirect (free radical) oxidation reactions. The following mass balance for the liquid phase in Figure 1 describes the reactor conditions assuming that substrate removal is by chemical oxidation only

$$Q_L \{[S_i]_o - [S_i]_e\} = V_L(k_i)_T [S_i]_e^m [Ox]^n \qquad (5)$$

Where
$$V_L = \text{Reactor volume, L}$$

If the overall oxidation reaction is assumed to be first order with respect to substrate concentration ($m = 1$), Equation (5) can be simplified and rearranged as Equation (6) which expresses the fraction of substrate remaining in solution if chemical oxidation is the only removal mechanism.

$$([S_i]_e / [S_i]_o)_{ox} = \frac{1}{[(HRT)(k_i)_T [Ox]^n + 1]} \qquad (6)$$

The overall simultaneous removal of substrate by the combined effects of the oxidation and volatilization mechanisms can be determined by combining Equations (3) and (6).

$$([S_i]_e / [S_i]_o)_T = \frac{1}{[(Q_g/Q_L)(H_i)(f(K_La)_i) + (HRT)(k_i)_T [Ox]^n + 1]} \qquad (7)$$

Equation (7) provides a general model for describing the removal of volatile and certain semi-volatile substrates under conditions which promote both volatilization and chemical oxidation removal mechanisms. The utility of Equation (7) however, is practically limited since measured values and/or functional relationships for critical variables such as $(k_i)_T$, $f(K_La)_i$, $[S_i]^m$ and $[Ox]^n$ are frequently unknown for the compounds of interest. Some data however, have been developed for ozonation of selected priority organic pollutant compounds[5,6]. These data can be combined with knowledge of mass transfer relationships for gas bubble-water contactors and ozone-substrate removal kinetics to demonstrate the applicability of Equation (7).

APPLICATION OF GENERAL MODEL TO OZONATION

Equation (7) can be modified for application to ozonation systems using bubble-contactors by providing appropriate expressions for $f(K_La)_i$, $[S_i]^m$ and $[Ox]^n$. Roberts et al[7] proposed the following relationship for $f(K_La)_i$ for mass transfer of compound "i" between water and diffused gas bubbles. It is applicable to ozone bubble contactors and to reactors with diffused aeration using liquid phase oxidants. It assumes that liquid-phase resistance controls the rate of mass transfer.

$$f(K_La)_i = 1 - \exp[-(K_La)_i (HRT)/(H_i (Q_g/Q_L))] \qquad (8)$$

Ozone oxidation kinetics have been shown to be second order overall but first order with respect to both substrate concentration and ozone concentration[8]. Thus, exponents m and n are unity in Equation (5) and Equation (7) is correct with n=1. Values of $(k_i)_T$ have not been developed for many of the organic priority pollutants. Hoigne and Bader[5,6] however, have developed values for the ozone depletion rate constant, $(k_i)_{O3}$, for many of these compounds in aqueous solutions. These data are presented in Tables I and II for selected compounds from USEPA's target compound lists for volatile organic (SW-8240) and semi-volatile organic compounds (SW-8270), respectively. Table III presents similar characterization data for other "non-listed" volatile and semi-volatile synthetic organics of environmental significance.

Terms $(k_i)_T$ and $(k_i)_{03}$ are related for any specific compound by the following relationship.

$$(k_i)_T = (k_i)_{03}/g \tag{9}$$

The constant "g" is the stoichiometric amount of ozone consumed by substrate oxidation and excludes the ozone demand of the background sample matrix. Values of g range from 1.0 for olefins to 2.5 for aromatic hydrocarbons. Review of the $(k_i)_{03}$ values in Tables 1 and 2 indicates that there is a large difference in their relative magnitudes. This difference would not be significantly effected by conversion of the measured $(k_i)_{03}$ values to $(k_i)_T$ values (if g were known for each compound). Therefore, the use of $(k_i)_{03}$ values, rather than $(k_i)_T$ values, in Equation (7) is reasonable and necessary.

Equation (7) can now be modified as follows for application to ozonation using bubble contact reactors.

$$([S_i]_e/[S_i]_o)_T = \frac{1}{[(Q_g/Q_L)(H_i)(1 - \exp(-q_i) + (HRT)(k_i)_T[O_3] + 1]} \tag{10}$$

Where

$$q_i = (K_La)_i (HRT)/(H_i(Q_g/Q_L)) \tag{10a}$$

TABLE I - SUMMARY OF OZONATION RATE CONSTANTS AND HENRY'S CONSTANTS FOR EPA TARGET COMPOUND LIST OF VOLATILE ORGANICS - SW-8240

Compound	O_3 Rate Constant[a] (L/m-s)	Henry's[b] Constant	Group No.
Carbon tetrachloride	<0.005	0.95	1
Chloroform	<0.1	0.16	1
Bromoform	<0.02	0.03	1
Methylene chloride	<0.1	0.12	1
Chlorobenzene	0.75	0.19	1
Benzene	2.0	0.22	1
Toluene	14	0.27	2
Ethylbenzene	14	0.27	2
o-Xylene	90	0.11	2
m-Xylene	94	0.10	2
p-Xylene	140	0.12	2
Tetrachloroethylene	<0.1	1.1	1
Trichloroethylene	17	0.48	2
cis-Dichloroethylene	<800	0.31	3
trans-Dichlroethylene	5,700	0.27	3
1,1-Dichloroethylene	110	7.8	2
Styrene	300,000	0.05	3
Acetone	0.032	0.001	1
Methyl ethyl ketone	0.07	0.002	1

[a]References (5) and (6).
[b]$H_i = H_m/RT$, dimensionless.

134

TABLE II - SUMMARY OF OZONATION RATE CONSTANTS AND HENRY'S CONSTANTS FOR EPA TARGET COMPOUND LIST OF SEMI-VOLATILE ORGANICS-SW-8270

Compound	O_3 Rate Constant[a] (L/m-s)	Henry's Constant[b]	Group No.
Nitrobenzene	0.09	5.0×10^{-4}	1
1,2,4-Trichlorobenzene	<1.6	0.06	1
1,4-Dichlorobenzene	<3	0.10	1
Benzoic acid	1.2[c]	0.002	1
Naphthalene	3,000	0.17	3
Phenol	1.8×10^7	1.3×10^{-5}	3
2-Chlorophenol	6.6×10^7	--	3
2,4-Dichlorophenol	5×10^9	--	3
2,4,6-Trichlorophenol	$>10^8$	--	3
2,4,5-Trichlorophenol	$>10^9$	--	3
Pentachlorophenol	$>10^5$	--	3
4-Nitrophenol	1.4×10^7	1.5×10^{-6}	3

[a]References (5) and (6).
[b]$H_i = H_m/RT$, dimensionless.
[c]Estimated from benzoate data.

TABLE III - SUMMARY OF OZONATION RATE CONSTANTS AND HENRY'S CONSTANTS FOR OTHER VOLATILE AND SEMI-VOLATILE ORGANICS

Compound	O_3 Rate Constant[a] (L/m-s)	Henry's Constant[b]	Group No.
Methanol	0.02	--	1
tert-Butanol	0.003	--	1
Ethanol	0.37	--	1
1-Propanol	0.37	--	1
1-Butanol	0.58	--	1
Formaldehyde	0.1	--	1
Tetrahydrofuran	<2[c]	--	1
Methyl isobutyl ketone	<30[c]	--	2
4-Chlorophenol	34×10^6	2.5×10^{-5}	3
1,2,3-Trimethylbenzene	400	0.24	3
1,3,5-Trimethylbenzene	700	0.24	3

[a]References (5) and (6).
[b]$H_i = H_m/RT$, dimensionless.
[c]Estimated from batch ozonation and stripping tests.

The value of $(K_La)_i$ in Equation (10a) was estimated from the relationship proposed by Roberts et al[7].

$$(K_La)_i = (Y_{i,02})/(Y_{03,02}) [(K_La)_{03}]$$ (11)

Where

$Y_{i,j}$ is the ratio of the mass transfer rate of compound "i" to that of "j" in a bubble-water system.

This relationship assumes that the mass transfer rates of volatile solutes (including oxygen and ozone) are proportional to one another. Data by Roberts et at[7] and Matter-Muller et al[9] indicate a mean value of $Y_{i,02}$ of 0.55 for several chlorinated aliphatics in laboratory grade water and secondary effluent. Gurol and Singer[10] reported a $Y_{03,02}$ value of 0.83. Substitution of these values into Equation (11) indicates that $(K_La)_i$ is approximately 65 percent of $(K_La)_{03}$. Although Equation (11) is strictly applicable for $H_i > 0.55$[11] it is reasonable for the purposes of this modeling analyses to use with H_i values of 0.19[7] and lower.

The total fractional substrate removal efficiency (FRE_T) for specific operating conditions is determined by Equation (11).

$$FRE_T = 1 - ([S_i]_e/[S_i]_o)_T$$ (12)

RESULTS AND ANALYSIS

The theoretical FRE_T was determined for a range of $(k_i)_T$ and H_i values assuming the following constant operating conditions and mass transfer rate. The $(K_La)_i$ value was based on an ozone mass transfer rate, $(K_La)_{03}$, of 60/hr.

$$Q_g/Q_L = 2.0$$
$$HRT = 15 \text{ minutes}$$
$$[O_3] = 1.0 \times 10^{-5} \text{ m/L}$$
$$(K_La)_i = 0.01/s$$

The results of the FRE_T analyses are presented in Figure 2 and are similar to the family of curves developed by Gurol[12]. Figure 2 indicates that compounds that are readily oxidized by ozone (e.g. napthalene with $(k_i)_T = 3,000$ L/m-s) have a high FRE_T even though they have negligible volatility ($H_i=0.02$). Trichloroethylene and tetra-chloroethylene however, are not readily oxidized by ozone and are primarily removed by volatilization at the assumed operating conditions.

The effects of process operational variables on compound removal performance were evaluated by dividing the volatile and semi-volatile compounds into three groups. The grouping is indicated in Tables I, II and III and were based on the magnitude of the compound's $(k_i)_T$ value. The characteristics of the groups' members are listed below. Many of the semi-volatile organics have extremely high $(k_i)_T$ values but negligible H_i values due to their low vapor pressure.

Group	$(k_i)_T$	H_i
1	< 10 L/m-s	$\leq 0.01 -- 0.95$
2	10 - 200 L/m-s	$\leq 0.1 - 7.8$
3	> 200 L/m-s	0.05 - 0.31

136

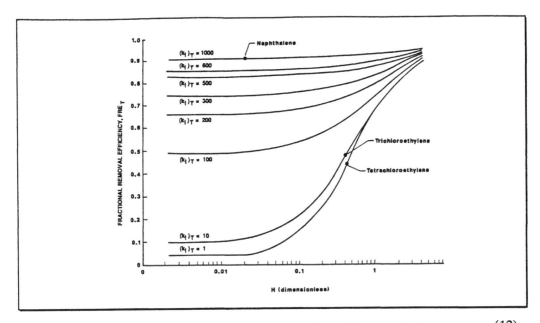

Figure 2- Fractional Removal Efficiency as a Function of H and $(ki)_T$ [After Gurol [12]]

EFFECT OF HYDRUALIC RETENTION TIME

The FRE_T was calculated for each of the compounds as a function of reactor hydraulic retention time (HRT) under constant operating conditions of $Q_g/Q_L = 2.0$, $O_3 = 4$ mg/L and $(K_L a)_i = 0.01/s$. The results for Groups 1, 2 and 3 of the volatile compounds are presented in Figures 3, 4 and 5, respectively and are discussed below.

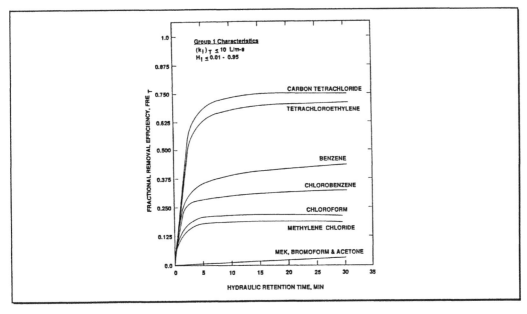

Figure 3 - Removal of Group 1 Volatile Organics by Ozonation and Volatilization with $Q_g/Q_L = 2.0$ and $O_3 = 4$ mg/L

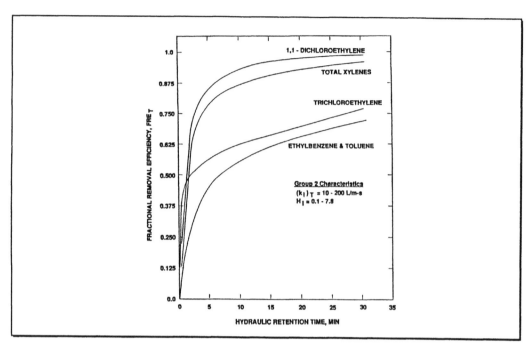

Figure 4 - Removal of Group 2 Volatile Organics by Ozonation and Volatilization with $Q_g/Q_L = 2.0$ and $O_3 = 4$ mg/L

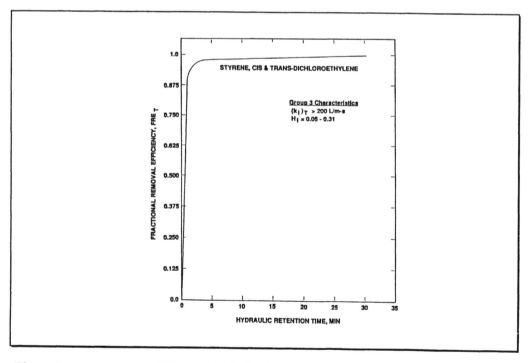

Figure 5 - Removal of Group 3 Volatile Organics by Ozonation and Volatilization with $Q_g/Q_L = 2.0$ and $O_3 = 4$ mg/l

All of the curves for Group 1 volatile compounds showed a rapid substrate removal in the first 2.0 to 2.5 minutes of the reaction. Thereafter, the curves became relatively flat and there was little additional substrate removal. Approximately 90 percent of the maximum FRE_T was achieved within the initial reaction period.

The substrate removal curves for Group 2 volatile compounds showed a similar rapid substrate removal during the initial reaction period. Unlike Group 1 however, Group 2 compounds demonstrate a continuous and significant substrate removal pattern with increasing HRT. Only one curve was needed to describe the Group 3 volatile compounds. It indicated a near instantaneous removal that achieved approximately 95 percent substrate reduction within a 2 minute reaction period.

Figure 6 was developed in order to differentiate and evaluate the relative effects of oxidation and volatilization on substrate removal. It presents the effect of HRT on the total fractional substrate removal and the fractional removal due to the oxidation mechanism (FRE_{ox}). The fractional substrate removal due to the volatilization mechanism, (FRE_v), is shown as the shaded area between the FRE_T and FRE_{ox} traces. Results are presented for chlorobenzene, total xylenes and cis-dichloroethylene as representative compounds from Groups 1, 2 and 3, respectively. The results presented in Figure 6 explain the consistent shapes of Group 1, 2 and 3 curves and provide insight to which process operational variables can be adjusted to improve the substrate removal performance of specific Groups. They indicate that essentially all of the observed removal of Group 1 substrates was due to rapid volatilization in the first 2 to 3 minutes. Thereafter, the FRE_T increased at a slow and constant rate due to the oxidation mechanism. In order to increase the FRE_T of Group 1 compounds, the amount of substrate oxidation must be increased through higher reactor ozone concentrations and/or addition of a catalyst.

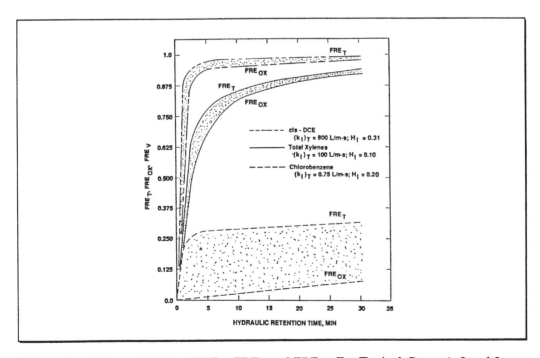

Figure 6 - Effect of HRT on FRE_T, FRE_v and FRE_{ox} For Typical Group 1, 2 and 3 Volatile Compounds with $Q_g/Q_L = 2.0$ and $O_3 = 4$ mg/L

Compounds in Groups 2 and 3 were removed predominantly by the oxidation mechanism. It appears therefore, that compounds having $(k_i)_T$ values greater than 10 L/m-s are readily removed by ozonation with negligible loss to volatilization under the assumed reactor operating conditions. Enhanced removal of Group 2 substrates can be most effectively achieved by increasing the reactor ozone concentration and/or its HRT. The removal of Group 3 compounds is strongly controlled by the oxidation mechanism regardless of the compound's volatility. Enhanced removal of Group 3 substrates is not practical or necessary. The removal rates of compounds having $(k_i)_T$ values greater than 10^4 L/m-s are probably controlled by mixing and diffusion rather than by the specific substrate reaction rate. All of the chlorinated phenols, nitrosophenols and phenol itself (Table 2 compounds) have $(k_i)_T > 10^5$ L/m-s.

EFFECT OF OZONE CONCENTRATION

The effect of reactor liquid phase ozone concentration on the removal efficiency of Group 1 and 2 volatile compounds is presented in Figure 7. All other reactor operating conditions were assumed constant with $Q_g/Q_L = 2.0$, HRT = 10 min and $(K_La)_i = 0.01/s$. For the purposes of developing Figure 7, it was assumed that increased reactor ozone levels could be acheived at a constant Q_g/Q_L ratio and $(K_La)_{03}$ value.

The results indicate that increased ozone concentration improved removal of Group 2 volatile substrates. Enhanced removal was particularly significant for toluene, ethylbenzene and trichloroethylene which have relatively low volatility ($0.27 < H_i < 0.48$). Removal of Group 1 compounds, however, was generally unaffected by increased reactor ozone concentration. Slight increases were observed in the FRE_T for chlorobenzene and benzene having $(k_i)_T$ values of 0.75 L/m-s and 2.0 L/m-s, respectively. It appears therefore, that the oxidation rate of the Group 1 volatile compounds is too slow to provide a significant increase in substrate removal by increased ozone concentration.

EFFECT OF SUBSTRATE MASS TRANSFER RATE

The effect of the substrate's mass transfer rate, $(K_La)_i$, on the removal efficiency of Group 1 volatile compounds was also investigated. The substrate removal effect was expressed by changes in the ozone mass transfer rate since they are proportionally related [Equation (11)] and since the $(K_La)_{03}$ term is a process design variable.

Increasing the ozone (and substrate) mass transfer rate increases the fractional removal by volatilization. Since stripping is already the predominant removal pathway for Group 1 volatile compounds, increasing $(K_La)_{03}$ only exacerbates the off-gas substrate concentration. Reducing $(K_La)_{03}$ to minimize volatilization however, is off-set by a lower liquid ozone concentration and/or higher gas flow rates. These conditions will actually increase the amount of substrate volatilized and reduce the amount that is oxidized. It appears therefore, that adjustment of the ozone mass transfer rate is not a significant process control parameter for removal of Group 1 volatile substrates.

EFFECTS OF PROCESS VARIABLES ON OZONATION OF SEMI-VOLATILE COMPOUNDS

The above analyses considered Group 1, 2 and 3 compounds in USEPA's target compound list of volatile organics as defined by SW-8240. Ozonation kinetic data were available for 19 of the 33 listed compounds. Less ozonation data are available however, for the target compound list of semi-volatiles as defined by SW-8270. Only 12 of the 66 listed target compounds have $(k_i)_T$ values and they are either Group 1 or Group 3 type substrates. The eight Group 3 substrates all have $(k_i)_T > 10^7$ L/m-s and are readily removed by the oxidation mechanism. The remaining four compounds are Group 1 substrates and have $(k_i)_T < 3$ L/m-s and H_i values of 0.0005 to 0.10.

The effects of HRT and liquid phase ozone concentration on the FRE_T of the Group 1 semi-volatile compounds are presented in Figures 8 and 9, respectively. Figure 8 was developed for varying reactor HRT values with constant operating conditions of $Q_g/Q_L = 2.0$, $O_3 = 4$ mg/L and $(K_La)_i = 0.01$/s. Figure 9 was developed for varying reactor ozone concentrations and constant operating conditions of $Q_g/Q_L = 2.0$, $(K_La)_i = 0.01$/s and HRT = 10 min. For purposes of developing Figure 9, it was assumed that increased reactor ozone levels could be achieved at a constant Q_g/Q_L ratio and $(K_La)_{O3}$ value.

The results indicated that nitrobenzene is not removed under any reasonable ozone contactor operating conditions. Benzoic acid and 1,2,4-trichloroethylene were removed at a maximum efficiency of 40 percent at HRT of 30 min and an O_3 concentration of 30 mg/L. The maximum FRE_T for 1,4-dichloroethylene was approximately 0.6 at HRT=30 min and an O_3 concentration of 30 mg/L.

SUMMARY AND CONCLUSIONS

This paper has developed a general mathematical model to predict the removal efficiency of volatile and semi-volatile organic compounds under reactor operating conditions that promote both chemical oxidation and volatilization of substrates. The general model is limited in utility since values and/or functional relationships for several kinetic and mass transfer coefficients are not available for many chemical oxidant systems. The necessary data however, have been developed for ozone oxidation of many synthetic organic compounds of environmental significance.

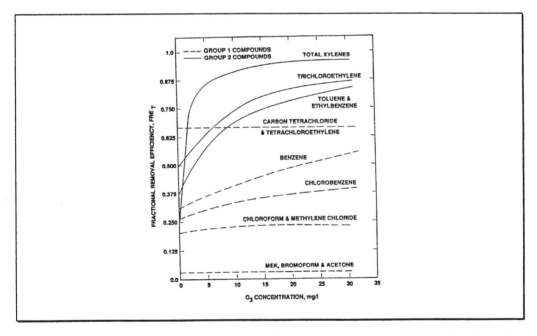

Figure 7 - Effect of Ozone Concentration on FRE_T Of Groups 1 and 2 Volatile Organics With HRT = 10 min and $Q_g/Q_L = 2.0$

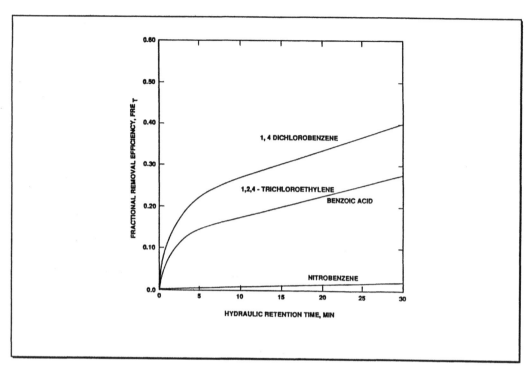

Figure 8 - Effect of HRT on FRE_T of Group 1 Semi-Volatile Compounds with $Q_g/Q_L = 2.0$ and $O_3 = 4$ mg/L

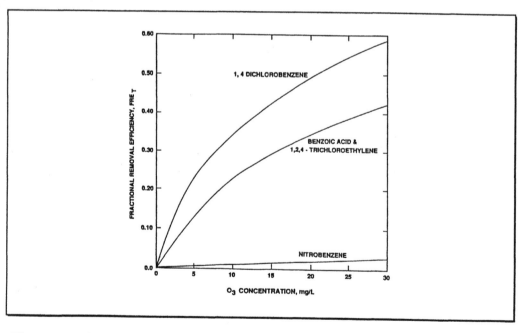

Figure 9 - Effect of O_3 Concentration on FRE_T Of Group 1 Semi-Volatile Compounds With $Q_g/Q_L = 2.0$ and HRT = 10 min

The general model was therefore modified to evaluate organic compound removal in ozonation systems that use gas bubble contactors. The revised model was used to evaluate the effects of hydraulic retention time, reactor ozone concentration and ozone mass transfer rate on removal of USEPA's target compound lists for volatile (SW-8240) and semi-volatile (SW-8270) compounds.

The results indicated that removal of Group 1 compounds $[(k_i)_T < 10$ L/m-s] was primarily by the volatilization mechanism since the ozone oxidation rate was too slow to exert a significant removal effect. Increasing the reactor ozone concentration and HRT did not significantly improve the FRE_T of Group 1 volatile or semi-volatile compounds. Increasing the substrate and ozone mass transfer rate increased the fraction of substrate removed by volatilization. These results indicate that enhanced removal of Group 1 volatile and semi-volatile compounds must be achieved by either addition of catalyst or by off-gas capture and recycle.

Group 2 compounds $(10 \leq (k_i)_T < 200$ L/m-s) are removed by a combination of rapid volatilization followed by gradual chemical oxidation. The FRE_T of these substrates significantly increased with increasing reactor HRT and ozone concentration. Group 2 compounds can be effectively removed by ozonation with a minimum off-gas substrate concentration.

Group 3 compounds $[(k_i)_T > 200$ L/m-s] were instantaneously removed by the oxidation mechanism even at low reactor ozone concentrations. Substrate removal by volatilization was not significant for the range of H_i values considered (1×10^{-5} to 0.3). Many of the listed semi-volatile organic compounds for which kinetic data were available have $(k_i)_T > 10^6$ L/m-s and can be effectively treated by chemical oxidation with ozone. There is however, limited ozonation kinetic data for the majority of the listed semi-volatile compounds.

REFERENCES

1. Barton, D.A., "Intermediate Transport of Organic Compounds in Biological Wastewater Treatment Processes". Environmental Progress, Vol. 6, No. 4, 1987, pg 246-256.

2. Berglund, R.L. and Whilpple, G.M., "Predictive Modeling of Organic Emissions." Chemical Engineering Progress, pg. 46-54, (November 1987).

3. Wang, L. et al, "Prediction of the Fate of Toxic Organic Compounds in Wastewater Treatment Systems". Presented at AIChE Summer National Meeting, Philadelphia, PA (1989).

4. Gurol, M.D. and Breman, W.M. , "Kinetics and Mechanism of Ozonation of Free Cyanide Species in Water". Environmental Science and Technology, Vol. 19, N 9, pg. 804(6) (September 1985).

5. Hoigne, J. and Bader, H., "Rate Constants of Reactions of Ozone With Organic and Inorganic Compounds in Water-I Non-Dissociating Organic Compounds. Water Research, Vol. 17, pg 173-183 (1983).

6. Hoigne, J. and Bader, H., "Rate Constants of Reactions of Ozone With Organic and Inorganic Compounds in Water - II Dissociating Organic Compounds. Water Research, Vol. 17, pg. 185-194 (1983).

7. Roberts, P.V. et al, "Modeling Volatile Organic Solute Removal by Surface and Bubble Aeration" Journal WPCF, Vol. 56, No. 2, pg. 157-163 (February, 1984).

8. Gurol, M.D. and Nekousinaini S., "Kinetic Behavior of Ozone in Aqueous Solutions of Substituted Phenols", Industrial and Chemical Engineering Fundamentals, Vol. 23, pg. 54 (January 1984).

9. Matter-Muller, C et al, "Transfer of Volatile Substances From Water to the Atmosphere" Water Research, Vol. 15, pg. 1271 (1981).

10. Gurol, M.D. and Singer P.C., "Kinetics of Ozone Decomposition: A Dynamic Approach". Environmental Science and Technology, Vol. 16, pg. 377 (1982).

11. Munz, C. and Roberts, P.V., "Gas-and Liquid-Phase Mass Transfer Resistances of Organic Compounds During Mechanical Surface Aeration", Water Research, Vol. 23, No. 5, pg. 589-601 (1989).

12. Gurol, M.D., "Factors Controlling the Removal of Organic Pollutants in Ozone Reactors". Journal AWWA, pg. 55-60 (August 1985).

Effects of Chemical Oxidation on Anaerobic Treatment of Phenols

ABSTRACT

The potential for using chemical oxidation to enhance anaerobic biodegradability and reduce toxicity of two model phenolic compounds (o-cresol and 2,4-DNP) was evaluated. Two types of bioassay were performed on the model compounds and their reaction products to determine sample biodegradability and toxicity in batch methanogenic cultures. Ozone, hydrogen peroxide, and potassium permanganate were the three oxidants examined in this study. Hydrogen peroxide, in the presence of a ferrous iron catalyst, was found to be the most efficient oxidant in both enhancing anaerobic biodegradability of o-cresol and reducing methanogenic toxicity of 2,4-DNP. A dose of hydrogen peroxide of about 1.5 g per g of o-cresol formed well biodegradable reaction products while less than 1 g of hydrogen peroxide per g of 2,4-DNP reduced methanogenic toxicity by 50%.

INTRODUCTION

Phenolic compounds represent the major constituents present in a wide spectrum of industrial wastewaters such as coal conversion, coke preparation, petroleum refineries, pulp and paper, and photoprocessing (1). Many of these compounds can cause adverse effects on human health and receiving waters if discharged without treatment. Effective treatment of these materials has been demonstrated by chemical oxidation using oxidants such as ozone (2-7), hydrogen peroxide (8-9), and permanganate (10). Chemical oxidation to achieve substantial reduction in chemical oxygen demand (COD) or total organic carbon (TOC), however, would be exceedingly costly and time consuming. Biological oxidation, however, is the most cost-effective treatment method for high-strength wastewaters. Chemical oxidation is therefore better suited for wastewaters containing compounds that are biorefractory or toxic to microorganisms.

Yi-Tin Wang, Assistant Professor, Department of Civil Engineering, University of Kentucky, Lexington, Kentucky, U.S.A.

Anaerobic biological treatment offers the advantages of no oxygen requirement, low waste biomass production, and the generation of methane gas. However, very little attention has been paid to assess the potential for using chemical oxidation to enhance anaerobic biological treatment of biorefractory or toxic materials. In this study, the effect of three oxidants on the anaerobic biodegradability and toxicity of two model phenolic compounds was examined in batch methanogenic cultures. Also, the relative effectiveness of three oxidants on enhancing biodegradation and reducing toxicity was evaluated.

MATERIALS AND METHODS

MODEL COMPOUNDS

Two phenolic compounds, o-cresol and 2,4-dinitrophenol (DNP) were used in this study. These compounds were selected because previous work has shown that they are biorefractory and/or inhibitory to methanogenic cultures (11-12). All phenolic solutions were prepared by adding predetermined amount of phenolic compounds to a proper volume of distilled and deionized water.

OZONATION

Ozone was generated from pure oxygen using a Welsbach Model T-816 Ozonator. Ozonation of model compounds was conducted in a 500 mL gas-washing bottle at an average feed gas ozone concentration of 0.23 mmol/min for a specific period of time to give the desired dosage. Ozonation was conducted with both unbuffered and buffered (at pH 9 with boric acid and sodium borate) solutions. pH was adjusted with HCl or NaOH.

HYDROGEN PEROXIDE OXIDATION

The reaction of phenols with hydrogen peroxide was conducted in batch stirred reactors. Ferrous iron ($FeCl_2$) was added as a catalyst for hydrogen peroxide oxidation at an optimum concentration ratio of phenolic compounds to iron of 100 to 1 and the reaction pH values were within the optimum range of 3 - 4 (8,9,13).

PERMANGANATE OXIDATION

Permanganate oxidation was conducted with both unbuffered and buffered solutions. The buffered solutions were prepared by adding 0.02 M boric acid and 0.02 M sodium borate and adjusting pH to 9 with NaOH.

All oxidant doses to batch reactors were expressed as g oxidant/g compound initially present.

BATCH BIOASSAY

The anaerobic biodegradability and toxicity of reaction products were evaluated by using two batch test procedures: the biochemical methane potential (BMP) and the anaerobic toxicity assay (ATA). In general, the procedures involved the placement of 50 mL sample and 50 mL microbial inoculum in a 125-mL serum bottle while under the purge of an oxygen-free gas mixture as described previously (14). The seed inoculum was a phenol-enriched methanogenic culture which had been maintained in our

146

laboratory for more than three years (15). Two types of substrates (acetic acid and phenol) were used in the ATA tests in evaluating the toxicity to the specific bacterial groups in mixed methanogenic cultures. Acetic acid was used to assess sample toxicity toward methanogenic bacteria while phenol was selected to estimate toxicity toward acetogenic bacteria in the phenol-enriched mixed culture.

Before being transferred to serum bottles, excess oxidants were first removed from all samples. Dissolved oxygen and residual ozone were removed by purging the samples with nitrogen gas for 15 min. Any excess hydrogen peroxide was removed by reacting with added ferrous iron (16). Residual permanganate was removed by first acidifying the samples to form MnO_2. The solid MnO_2 was then removed from solutions by filtering the samples through a 0.45 μm membrane filter.

All samples were buffered at pH 7.0 with sodium bicarbonate before finally sealed with butyl rubber stoppers and placed on a continuous shaker in a 35°C constant temperature room. Duplicates were run for all samples, including the controls.

ANALYTICAL PROCEDURE

The ozone concentration in the feed gas was determined by an iodometric method (17). Gas production was monitored by displacement of the plunger in an appropriate sized wetted glass syringe. Gas composition was determined with a gas partitioner (Fisher Model 1200) using certified calibration standards. COD was determined by using Method 508 of Standard Methods (17). TOC was measured by chemical ultraviolet oxidation and infrared detection using a carbon analyzer (Dohrman Model DC-80). o-Cresol and 2,4-DNP were analyzed by a liquid chromatograph (Varian Model 5020) with a MicroPak MCH-5 C_{18} reverse phase column at ambient temperature. Organic acids were analyzed by measuring their butyl esters with a gas chromatograph (Varian Model 3400) (15).

RESULTS AND DISCUSSION

The anaerobic biodegradability of reaction products of unbuffered 200 mg/L o-cresol and 100 mg/L 2,4-DNP under a range of oxidant doses was shown in Figure 1. Biodegradability of reaction products was indicated by the ratio of the actual and the potential methane production. The actual methane production was determined by substracting methane produced in the seed-blank bottles from that measured in test bottles. Experimental errors of determining actual methane production were estimated to be within ±10%. The methane potential was calculated based on the measured COD of the sample and a conversion factor of 0.35 L CH_4/g COD at 0°C and 1 atm (18). The data in Figure 1 illustrate that the biodegradability of model compounds increased with increasing doses of oxidants. The rate of increase was the greatest with hydrogen peroxide and followed by permanganate and ozone. For 1 g of o-cresol, only about 1.5 g hydrogen peroxide was needed to obtain a methane production of 40% of its potential while about 10 g permanganate and 12 g ozone were needed to achieve the same biodegradability, respectively. In the case of 2,4-DNP, oxidation was less effective to generate biodegradable products. The oxidation products of 2,4-DNP were much less biodegradable than those of o-cresol under the same dose of ozone. Furthermore, the reaction products of 2,4-DNP with either hydrogen peroxide or permanganate were not significantly degraded as no actual methane production was noted in any of these sample bottles.

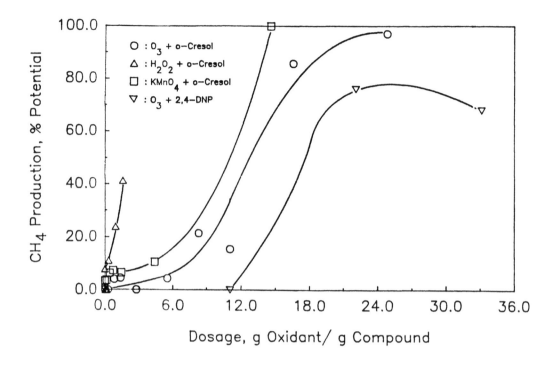

Figure 1. Anaerobic Biodegradability of Reaction Products

The observed rate of increase in biodegradability of o-cresol was not correlated to the rate of removal of initial compounds, instead to the rate of TOC reduction by these oxidants (Tables 1 and 2 and Figure 2). Hydrogen peroxide was the most efficient oxidant in removing the two phenols as well as in reducing the TOC. The strong oxidation power of hydrogen peroxide is attributed to the formation of hydroxyl free radicals in the presence of a ferrous iron salt (9). The data in Tables 1 and 2 indicate that ozone was less efficient than hydrogen peroxide in removing phenols in acid solutions. During the course of ozonation, the pH of the unbuffered phenolic solutions quickly dropped to about 3. Ozone is a stronger oxidant in basic solutions due to formation of highly reactive hydroxyl radicals from ozone decomposition. The observed rapid pH drops corresponded to formation of organic acids. Ozonation of o-cresol and 2,4-DNP produced formic acid, acetic acid, propionic acid, glyoxylic acid and oxalic acid and these acids accounted for 21 - 75% TOC present after complete removal of initial compounds. Formic acid, acetic acid, and oxalic acid were also detected during hydrogen peroxide oxidation of o-cresol and these acids accounted for about 18% measured TOC after complete disappearance of o-cresol. The permanganate reacted samples were not analyzed for organic acids. However, formation of the organic acids was indicated by pH drops from 3.4 to 2.7 in the unbuffered solutions (Table 2). BMP tests conducted on the identified acids revealed that formic acid, acetic acid, and propionic acid were biodegradable to methane while oxalic acid and glyoxylic acid were not significant degradaded in the phenol-enriched methanogenic culture.

TABLE 1. SUMMARY OF CHEMICAL OXIDATION OF O-CRESOL

O-Cresol concentration (initial, mg/l)	Oxidant and dose (g/g o-cresol)	End pH at highest dose	Oxidant needed for 100% removal (g/g o-cresol)	Product identified
200	O_3 : 0.5-24.8	2.7	1.7	Formic, acetic,
600	O_3 : 0.5-22.1	2.5	2.2	propionic, glyoxylic,
600	O_3 : 0.5-22.1	8.3	1.8	oxalic and salicylic acid
200	H_2O_2: 0.03-3.15	3.1	1.2	Formic, acetic, and
600	H_2O_2: 0.03-3.15	2.8	1.6	oxalic acid
200	$KMnO_4$: 0.09-8.6	3.3	4.4	——
600	$KMnO_4$: 0.09-8.6	4.7	——	

TABLE 2. SUMMARY OF CHEMICAL OXIDATION OF 2,4-DNP

2,4-DNP concentration (initial, mg/l)	Oxidant and dose (g/g 2,4-DNP)	End pH at highest dose	Oxidant needed for 100% removal (g/g 2,4-DNP)	Product identified
100	O_3 : 2.8-49.7	2.8	5.5	Formic, acetic,
100	O_3 : 2.8-49.7	8.3	3.9	glyoxylic, and oxalic acid
100	H_2O_2 : 0.02-0.9	2.8	0.9	——
100	$KMnO_4$: 0.09-8.6	2.7	>4.3	——
100	$KMnO_4$: 0.09-8.6	8.7	4.3	

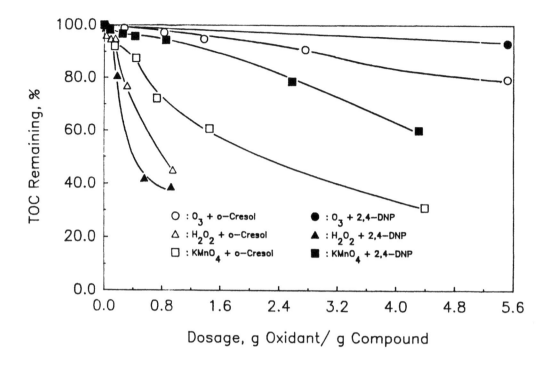

Figure 2. Reduction of TOC During Chemical Oxidation

However, no methane production was observed in any of the samples containing reaction products of 2,4-DNP reacted with either hydrogen peroxide or permanganate, indicating that biodegradation of biodegradable organic acids was inhibited by unidentified products. Although permanganate removed 2,4-DNP much faster in the basic solutions than in the acid solutions, no methane production was observed with reaction products formed in solutions buffered at pH 9 even under a high dose of 8.6 g $KMnO_4$/g 2,4-DNP. Inhibitory products formed during ozonation of 2,4-DNP, however, may be only transient intermediates since sufficient doses yielded a significant fraction of biodegradable products. Thus, the rate of formation of biodegradable products from o-cresol was related to the rate of TOC removal by these oxidants. In addition, large doses of oxidants may or may not improve the biodegradability of 2,4-DNP due to formation of stable and inhibitory products.

The ATA tests were conducted to examine the toxic or inhibitory nature of the reaction products. Figures 3 and 4 show the cumulative methane production in sample as well as control bottles with 600 mg/L o-cresol as the initial compound and permanganate as the oxidant. Because the degradation of phenol to methane requires the cooperation of phenol-degrading bacteria and acetate-utilizing methanogenic bacteria, both acetic acid- and phenol-supplemented cultures were used in the ATA tests. Inhibition to the acetate-utilizing methanogens was indicated by a

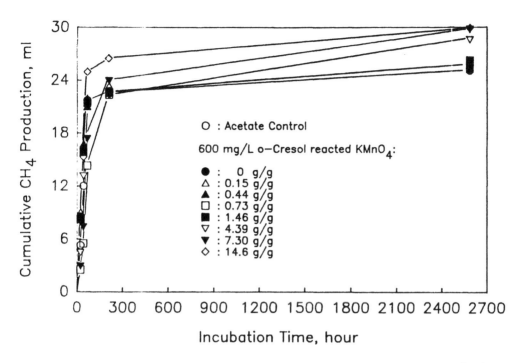

Figure 3. Methane Production from Acetate-Spiked Reaction Products

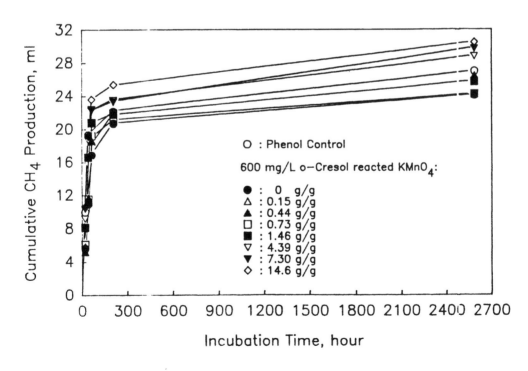

Figure 4. Methane Production from Phenol-Spiked Reaction Products

decreased rate of methane production relative to the acetate control while inhibition with phenol degradation was indicated by a decreased rate of methane production relative to the phenol control. The acetate control received only 500 mg/L acetate, whereas the phenol control received only 200 mg/L phenol. The data in Figures 3 and 4 indicate that o-cresol and its reaction products formed under lower doses of permanganate were slightly inhibitory to both the acetate-utilizing methanogens and the phenol degraders as the rates of methane production were lower than those of controls. However, methane production in sample bottles at the end of incubation approximated or even exceeded those produced in the control bottles, indicating acclimation or biodegradation of reaction products. A similar pattern was also noted for o-cresol with ozone and hydrogen peroxide as oxidants (data not shown).

For 2,4-DNP, the initial compound was more inhibitory than its reaction products and a progressive decrease in inhibition was observed with increasing doses of either ozone or hydrogen peroxide. An example of results from ATA tests performed on ozonation products with phenol-supplemented cultures are presented in Figure 5. No significant reduction in toxicity was observed for both buffered (at pH 9) and unbuffered 100 mg/L 2,4-DNP reacted with permanganate. The reduction of toxicity was quantified by defining relative toxicity, T_r, as

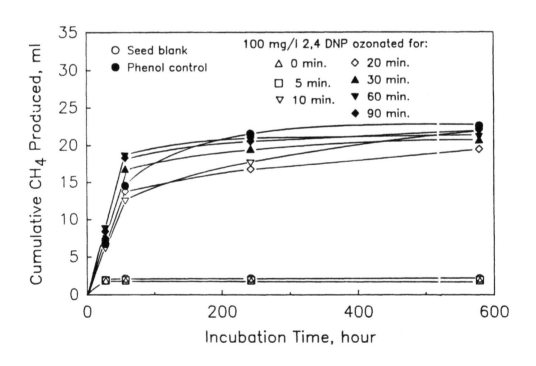

Figure 5. Methane Production from Phenol-Spiked Ozonation Products: 2,4-DNP Ozonated at pH 3

$$T_r = T/T_o \qquad (1)$$

where

    T = toxicity of samples to methane production

    = $1 - v/V$

    T_o = toxicity of initial compound to methane production

    = $1 - v_o/V$

where

    v = total volume of methane produced from samples, mL

    v_o = total volume of methane produced from initial compound, mL, and

    V = total volume of methane produced from control, mL.

Figures 6 and 7 show the toxicity of reaction products relative to the initial compound, 2,4-DNP. The reduction in toxicity by ozonation did not differ significantly between the acetate-utilizing methanogens and the phenol-degrading bacteria nor was it affected appreciably by the pH during the ozonation reaction. However, the rate of toxicity reduction by hydrogen peroxide was faster for acetate-utilizing methanogens than phenol-degraders, indicating that phenol-degrading bacteria were more susceptible to inhibition of reaction products than acetate-utilizing methanogens (Figure 7). In addition, Figures 6 and 7 show that hydrogen

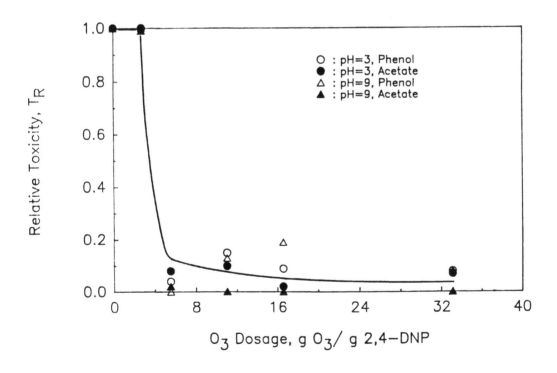

Figure 6. Effect of Ozonation on the Toxicity of 2,4-DNP

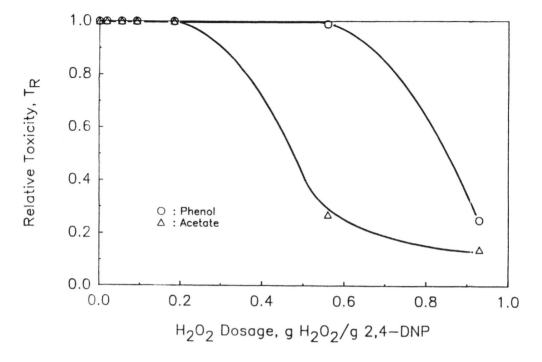

Figure 7. Effect of Hydrogen Peroxide Oxidation on the Toxicity of 2,4-DNP

peroxide was more efficient in reducing toxicity than ozone. The doses needed for 50% toxicity reduction were less than 1 g for hydrogen peroxide and about 3 g for ozone per g of 2,4-DNP. However, no toxicity reduction was noted with permanganate even under a high dose of 8.6 g/g 2,4-DNP.

CONCLUSIONS

Chemical oxidation enhances the anaerobic biodegradability of o-cresol by forming biodegradable products. Chemical oxidation also reduces the toxicity of 2,4-DNP to a phenol-enriched methanogenic culture. Hydrogen peroxide, in the presence of a ferrous iron catalyst, is a more effective oxidant in both generating biodegradable products and reducing methanogenic toxicity than ozone and permanganate.

ACKNOWLEDGMENTS

The author wishes to thank James L. Latchaw and Pin-Chieh Pai for conducting the laboratory work.

REFERENCE

1. Patterson, J.W. Wastewater Treatment Technology. Ann Arbor: Ann Arbor Science, 1975.

2. Eisenhauer, H.R. The Ozonation of Phenolic Wastes. J. Water Pollu. Control Fed., 40: 1887, 1968.

3. Gould, J.P., and Weber, W.J. Oxidation of Phenols by Ozone.J. Water Pollu. Control Fed., 48: 47, 1976.

4. Neufeld, R.D., and Spinola, A.A. Ozonation of Coal Gasification Plant Wastewater. Environ. Sci. Technol., 12: 470, 1978.

5. Singer, P.C., and Gurol, M.D. Dynamics of the Ozonation of Phenol - I. Experimental Observations. Water Res.,17: 1163, 1983.

6. Yamamoto, Y., Niki, E., Shiokawa, H., and Kamiya, Y. Ozonation of Organic Compounds.II. Ozonation of Phenol in Water. J. Org. Chem., 44: 2137, 1979.

7. Baillod, C.R., Faith, B.M., and Masi, O. Fate of Specific Pollutants During Wet Oxidation and Ozonation. Environ. Prog., 1: 217, 1982.

8. Keating, E.J., Brown, R.A., and Greenberg, E.S. Phenolic Problems Solved with Hydrogen Peroxide. Proc. 33rd Purdue Ind. Waste Conf., p471, 1979.

9. Eisenhauer, H.R. Oxidation of Phenolic Wastes: I. Oxidation with Hydrogen Peroxide and a Ferrous Salt Reagent. J. Water Pollu. Contr. Fed., 36: 1116, 1964.

10. Lee, D.G., and Sabastian, C.F. The Oxidation of Phenol and Chlorophenols by Alkaline Permanganate. Canadian J. Chem., 59: 2776, 1981.

11. Wang, Y. Y., Suidan, M.T., Pfeffer, J.T., and Najm, I. Effects of Some Alkyl Phenols on the Methanogenic Degradation of Phenol. Appl. Environ. Microbiol., 54: 1277, 1988.

12. Boyd, S.A., Shelton, D.R., Berry,D., and Teidje, J.M. Anaerobic Biodegradation of Phenolic Compounds in Digested Sludge. App. Environ. Microbiol.,46: 50, 1983.

13. Bowers, A. R., Gaddipati, P., Eckenfelder, W. W., and Monsen, R.M. Treatment of Toxic or Refractory Wastewaters with Hydrogen Peroxide. Wat. Sci. Tech., 21: 477, 1989.

14. Wang, Y.T. Methanogenic Degradation of Ozonation Products of Biorefractory or Toxic Aromatic Compounds. Wat. Res., 24: 185, 1990.

15. Wang, Y.T., Pai, P.C., and Latchaw, J.L. Effects of Preozonation on Anaerobic Biodegradability of o-Cresol. J. Environ. Eng.,115: 336, 1989.

16. Wang, Y.T., and Latchaw, J.L. Anaerobic Biodegradability and Toxicity of Hydrogen Peroxide Oxidation Products of Phenols. Res. J. Wat. Pollu. Cont. Fed., 62: 234, 1990.

17. Am. Public Health Assoc. Standard Methods for the Examination of Water and Wastewater. 16th ed., Washington, D.C., 1985.

18. Wang, Y.T., Pai, P.C., and Latchaw, J.L. Methanogenic Toxicity Reduction of 2,4-Dinitrophenol by Ozone. Hazardous Waste & Hazardous Materials, 6: 33, 1989.

GILBERT GORDON

Redox Reactions and the Analytical Chemistry of Chlorine Dioxide and Ozone

ABSTRACT

Ozone and chlorine dioxide are strong oxidants which are being given considerable attention as alternative oxidants for the disinfection of drinking water. The details of the inorganic oxidation-reduction (redox) reactions of both ozone and chlorine dioxide have been described in detail in the literature for many years. Only recently has the mechanism of the self-decomposition of ozone in aqueous solution become reasonably well understood. It is becoming clear that ozone can be involved in both one and two electron transfer redox processes.

On the other hand, chlorine dioxide is most frequently encountered as a one-electron oxidant. The formation of highly reactive, complex intermediates is commonplace in the reactions of chlorine dioxide and chlorite ion. Their role is also of primary importance in understanding the microscopic details of most oscillating reactions where one- and two-electron reactions play a pivotal role. This paper will review the inorganic chemistry of ozone and chlorine dioxide along with the associated underlying by-product chemistry of both species -- with emphasis on the role of highly reactive non-metal species in these processes.

Existing analytical methods for ozone and chlorine dioxide also will be reviewed. Some of the methods reviewed require only modest additional field/laboratory evaluation before they could be designated as recommended methods. On the other hand some of the newly published methods are in such a preliminary state of development that they will require many months/years before they can be demonstrated and recommended as advantageous methods. Specifically, guidance and recommendations in selecting residual monitoring techniques for ozone, chlorine dioxide, chlorine/hypochlorous (free chlorine), chlorite ion and chlorate ion will be provided.

Gilbert Gordon, Miami University, Department of Chemistry, 124 Hughes Laboratories, Oxford, Ohio, U.S.A.

INTRODUCTION

Ozone and chlorine dioxide are strong oxidants which continue to receive considerable attention as alternative disinfectants for drinking water. The inorganic oxidation-reduction (redox) reactions of both ozone and chlorine dioxide have been described in detail in the literature for many years (1-2). Only recently has the mechanism of the self-decomposition of ozone in aqueous solution become reasonably well understood (3). It is becoming clear that ozone is involved in both one and two electron transfer processes. However, chlorine dioxide is most frequently encountered as a one-electron oxidant. The formation of highly reactive, complex intermediates is commonplace in the reactions of chlorine dioxide and chlorite ion. Their role is also of primary importance in understanding the microscopic details of most oscillating reactions where one- and two-electron reactions and highly reactive metal ion species play a pivotal role (4).

One of the difficulties in understanding these reactions is our inability to discriminate among the various species that are present when ozone or chlorine dioxide are used as oxidants. For example, when chlorine dioxide is used, not only must chlorine dioxide be measured, but also the species used to generate chlorine dioxide, such as free chlorine and chlorite ion, and the related by-products, chlorite ion and chlorate ion must be measured. Indirect measurement of these species is not desirable. Speciation can be accomplished either by using chemical masks to eliminate interferences and/or by taking advantage of the kinetic rate differences under automated conditions.

Any method which determines concentrations by difference is potentially inaccurate and subject to large accumulative errors -- both in terms of accuracy and precision. The subtraction of two large numbers to produce a small number means that the errors associated with those large numbers are propagated to the small number. The result in many cases is that the error is larger than the smaller number, therefore, giving meaningless information. Methods which obtain values by differences should be avoided.

Increased chemical knowledge about ozone and its decomposition along with chlorine dioxide and its by-products, chlorite ion and chlorate ion not only benefit the analytical methodology, but also benefit our understanding of the closely related redox reactions.

Chlorine Dioxide. Chlorine dioxide readily dissolves in water to form a solution which is biocidal to a wide range of microorganisms (5). Several studies have been conducted in an effort to determine if there are any detrimental health effects caused by chlorine dioxide and its metabolites and/or decomposition products, chlorite ion and chlorate ion (6-8). It does appear that the real species of concern are chlorite ion and chlorate ion.

In other words, one of the major reason for not using high concentrations of chlorine dioxide in the treatment of drinking water is that the ClO_2^- and ClO_3^- by-products are potential health risks. If it were possible to safely reduce or eliminate these by-products at low cost, chlorine dioxide well might become the disinfectant of choice.

Chlorine dioxide can be generated using several reaction schemes such as the reaction of aqueous hypochlorous acid with dissolved chlorite ion:

$$2HClO_2 + HOCl \rightarrow HCl + H_2O + 2ClO_2$$

Chlorine dioxide can also be generated by the direct reaction of sodium chlorite in the solid state, in solution with a mineral acid, with chlorine or with hypochlorous acid. The reaction for chlorine and/or hypochlorous acid (9-12) with chlorite ion is:

$$2ClO_2^- + Cl_2(g) \rightarrow 2ClO_2(g) + 2Cl^- \tag{1}$$

$$2ClO_2^- + HOCl \rightarrow 2ClO_2(g) + Cl^- + OH^- \tag{2}$$

These reactions involve the formation (20-22) of the unsymmetrical redox intermediate Cl_2O_2:

$$Cl_2 + ClO_2^- \rightarrow [Cl_2O_2] + Cl^- \tag{3}$$

At high concentrations of both reactants, the intermediate is formed very rapidly. Chlorine formed in Eqn.4 is recycled by means of Eqn. 3. Thus, mainly chlorine dioxide is produced as a result:

$$2[Cl_2O_2] \rightarrow 2ClO_2 + Cl_2 \tag{4}$$

(or) $$[Cl_2O_2] + ClO_2^- \rightarrow 2ClO_2 + Cl^- \tag{5}$$

On the other hand, at low initial reactant concentrations or in the presence of excess chlorine or hypochlorous acid (free available chlorine), primarily chlorate ion is formed due to the following reactions:

$$[Cl_2O_2] + H_2O \rightarrow ClO_3^- + Cl^- + 2H^+ \tag{6}$$

and $$[Cl_2O_2] + HOCl \rightarrow ClO_3^- + Cl^- + H^+ \tag{7}$$

Thus, during the generation process, high concentrations of excess chlorite ion favor the second order reactions (Eqns. 4-5) and chlorine dioxide is formed. At low concentrations, the second order disproportionation process becomes unimportant and reactions 6 and 7 produce chlorate ion rather than chlorine dioxide. The reasons for the production of chlorate ion are related to presence of high concentrations of free chlorine and the rapid formation of the Cl_2O_2 intermediate which in turn reacts with the excess hypochlorous acid to form the unwanted chlorate ion.

The stoichiometry of the undesirable reactions which form chlorate ion is:

$$ClO_2^- + HOCl \rightarrow ClO_3^- + Cl^- + H^+ \tag{8}$$

$$ClO_2^- + Cl_2 + H_2O \rightarrow ClO_3^- + 2Cl^- + 2H^+ \tag{9}$$

The most effective way to minimize chlorate ion formation is to avoid conditions which result in low reaction rates (e.g. high pH values and/or low initial reactant concentrations and the presence of free available chlorine). Clearly, the chlorate ion forming reaction (Eqn 6) will be more troublesome in dilute solutions. On the other hand, whenever treatment by chlorine dioxide (which forms free chlorite ion in the process) is followed by the addition of free chlorine (actually HOCl in the ph 5-8 region), the unwanted chlorate ion will also be formed.

In many older chlorine dioxide generators, hydrochloric acid is mixed with the chlorine solution before reaction with the sodium chlorite. The acid shifts the chlorine solution equilibria (the hypochlorous acid dissociation, Eqn 10 and the chlorine hydrolysis equilibrium, Eqn 11) favoring molecular chlorine (13-14).

$$HOCl = OCl^- + H^+ \tag{10}$$

$$Cl_2 + H_2O = HOCl + H^+ + Cl^- \tag{11}$$

When the pH is maintained between 2 and 3, yields of more than 90 percent ClO_2 can be achieved. Higher or lower pH values result in decreased yields with some excess chlorine remaining in the chlorine dioxide solution (5).

One of the most exciting recent developments in chlorine dioxide generator technology utilizes the reaction of chlorine with a concentrated sodium chlorite solution under vacuum (5, 13-14) resulting in yields of chlorine dioxide solutions in excess of 95 percent with less than 5 percent excess chlorine -- with the minimum formation of any chlorate ion. However, in this context it should be noted that almost all old chlorine dioxide generators now in use in water treatment plants throughout the United States continue to produce some chlorate ion unless they are very carefully adjusted and properly monitored.

It should be clear from the preceding discussion that the generation and decomposition of chlorine dioxide may produce chlorine and/or hypochlorous acid, chlorite ion, chlorate ion, and chloride ion.

ANALYTICAL METHODOLOGY. The coexistence of all the reactants, product and by-product species means that without appropriate precautions many existing analytical methods measure the sum total of chlorine dioxide, chlorine, and chlorite ion.

Without question the most important parameter of an analytical method is accuracy. To achieve good accuracy it is implied that the method be selective. One way to achieve this selectivity is to use a specific or selective reagent. Unfortunately, these types of reagents are not common, therefore other approaches to achieve selectivity, such as kinetic based methods, have been applied. Flow injection analysis (FIA) allows analysts to automate kinetic based analytical measurements.

The introduction of FIA, makes the reproducible handling of solutions with strict control of the timing possible using simple instrumentation (15-17). Two fundamental FIA approaches which utilize differences in rates are kinetic discrimination and kinetic enhancement (16). In kinetic discrimination the reaction rate differences between the reagent-analyte and reagent-interferences are optimized and the enhanced selectivity is utilized to create a new analytical method. In kinetic enhancement, the reaction rate for the analyte is maximized thereby creating a method with lower limits of detection. Currently, the FIA procedure and ion chromatography are being compared in order to understand the advantages and disadvantages of each technique. Both methods will be published by the EPA in early 1991. All evidence would suggest that especially for low level analysis of chlorite ion and chlorate ion (< 1 mg/L), the methods are complimentary.

With mildly basic conditions (pH 7-8.5) chlorine dioxide reacts with iodide ion to form iodine and chlorite ion, while chlorine reacts with iodide ion to form iodine and chloride ion.

$$Cl_2 + 2I^- = I_2 + 2Cl^- \qquad \text{(pH below 8.5)} \qquad (12)$$

$$2ClO_2 + 2I^- \rightarrow I_2 + 2ClO_2^- \qquad (13)$$

Under rather acidic conditions (pH 2 or less) chlorite ion reacts with iodide ion to form iodine and the chlorite ion is reduced to chloride ion. With these conditions, chlorine dioxide reacts with iodide ion to give five moles of iodine and 2 moles of chloride ion.

$$2ClO_2 + 10I^- + 8H^+ = 5I_2 + 2Cl^- + 4H_2O \quad \text{(pH} \leq 2) \qquad (14)$$

$$ClO_2^- + 4I^- + 4H^+ \rightarrow 2I_2 + Cl^- + 2H_2O \qquad (15)$$

The reaction of chlorate and iodide ions under these conditions is very slow, and requires severe acid conditions (>6M HCl) for the reaction to be of use analytically:

$$ClO_3^- + 6I^- + 6H^+ \rightarrow 3I_2 + Cl^- + 3H_2O \qquad (16)$$

Mixtures of chlorite and chlorate ions in the absence of chlorine dioxide and/or chlorine are determined sequentially. In this case, chlorite ion is mixed with 0.02 M hydrochloric acid and 0.3 M potassium iodide which is used to form I_3^- and to maintain constant iodide ion concentration. The sample is mixed with the reagent as they flow through the tubing to the detector. This is followed by the determination of both chlorite ion and chlorate ion using 12 M hydrochloric acid. The calibration curves for the mixtures are linear in the 0.02 - 2.0 mg/L range.

Chlorine and chlorine dioxide are determined iodometrically at pH 8.3 without interference from chlorite or chlorate ions by modification of the standard FIA setup. Independent calibration curves show linear ranges of 0.2 - 10 mg/l for chlorine and 0.3 - 10 mg/l for chlorine dioxide with a correlation coefficient of 0.999 or better. The corresponding uncertainties for the chlorine and chlorine dioxide determination are ± 0.01 mg/l at concentrations of 1.50 mg/l and ± 0.02 mg/l at 0.3 mg/l chlorine or chlorine dioxide.

For the HOCl-HClO$_2$ reaction that produces both chlorine dioxide and chlorate ion, it is not clear whether the two products are formed concurrently or independently.

$$2ClO_2 + H_2O \rightarrow ClO_3^- + ClO_2^- + 2H^+ \qquad (17)$$

$$2ClO_2 + HOCl + H_2O \rightarrow 2ClO_3^- + Cl^- + 3H^+ \qquad (18)$$

The reaction is not catalyzed by metal ions and the addition of HgCl$_2$ does not alter the rate of formation of ClO$_3^-$ at pH 7.87. Thus, the kinetic results are interpreted in terms of the following mechanism:

$$HOCl + HClO_2 \underset{k_{-1}}{\overset{k_1}{=}} Cl_2O_2 + H_2O \qquad (19)$$

$$Cl_2O_2 + HOCl \overset{k_2}{\rightarrow} ClO_3^- + Cl_2 + H^+ \qquad (20)$$

where Cl$_2$O$_2$ is the same intermediate as was proposed by Taube and Dodgen (18) and others (8-10,16-17). This corresponds to the stoichiometric equation:

$$2\ HOCl + HClO_2 = ClO_3^- + Cl_2 + H_3O^+ \qquad (21)$$

In past, the various reactions and interactions of the oxy-halogen species were not well documented in the literature. However, the reactions of chlorine and hypochlorous acid have received considerable attention because of their long use as disinfectants in North America. Recently, the reactions of chlorine dioxide, chlorite ion, bromate ion and other closely associated species have become of interest because of the ease with which they undergo electron transfer/oxidation reduction reactions. In general, many of these details appear in the literature under the classification of oscillation reaction (4). Specifically, these reactions involve a variety of highly reactive oxyhalogen species such as Cl$_2$O$_2$, HOCl, Cl$_2$O, ClO$_2^-$, and BrO$_2^-$.

OZONE. Many early attempts were made to study the decomposition of aqueous ozone and to determine the mechanistic role and reactions of its intermediate decomposition products. Various kinetic orders (and combinations of orders) ranging from 0.5 to 2 have been proposed (3, 19-22). Unfortunately, many of the studies were not conducted under comparable conditions (i.e., different ozone concentrations, pH, and ionic media, buffers present or absent, possible scavengers and promoters present, etc.). These variations in experimental conditions have produced "system-specific" rate constants (3).

The early works of Taube and Bray (23) and others (19, 22), however, clearly point out the chain-reaction characteristics of the decomposition products of aqueous ozone, and note the role of various scavengers such as formic acid, bromide ion, and acetic acid. Recent papers from the laboratories of Gordon (3), Hoigné (24-26), and Hart (27-28) take advantage of free radical scavengers such as acetic acid, carbonate ion, phosphate ion, and t-butyl alcohol to define specific systems allowing for the elucidation of the various propagation and termination steps associated with the decomposition of aqueous ozone.

Clearly, the decomposition of ozone involves the very reactive and catalytic intermediates O_2^-, O_3^-, OH, HO_2, HO_2^- and H_2O_2. In turn, these species markedly influence the stability of ozone solutions and readily alter the precision and accuracy of various analytical methods for dissolved ozone. These species also influence the role and magnitude of various potential interferents on the ozone decomposition process.

In this context, it should be noted that the stability of dissolved ozone is readily affected by pH, UV light, concentration of ozone, and the concentration of radical scavengers (3, 26). The role of various free radicals now is recognized as incontrovertible.

The most widely tested and accepted rate law for the decomposition of aqueous ozone is consistent with the following mechanistic steps (Eqns 22-32):

initiation:

$$O_3 + OH^- \rightarrow HO_2^- + O_2 \qquad\qquad 40 \text{ M}^{-1}\text{s}^{-1} \qquad\qquad (22)$$

propagation/termination:

$$HO_2^- + O_3 \rightarrow O_3^- + HO_2 \qquad 2.2 \times 10^6 \text{ M}^{-1}\text{s}^{-1} \qquad (23)$$

$$HO_2 + OH^- \rightarrow O_2^- + H_2O \qquad pK = 4.8 \qquad (24)$$

$$O_2^- + O_3 \rightarrow O_3^- + O_2 \qquad 1.6 \times 10^9 \text{ M}^{-1}\text{s}^{-1} \qquad (25)$$

$$O_3^- + H_2O \rightarrow OH + O_2 + OH^- \qquad 20\text{-}30 \text{ M}^{-1}\text{s}^{-1} \qquad (26)$$

$$O_3^- + OH \rightarrow O_2^- + HO_2 \qquad 6 \times 10^9 \text{ M}^{-1}\text{s}^{-1} \qquad (27)$$

$$O_3^- + OH \rightarrow O_3 + OH^- \qquad 2.5 \times 10^9 \text{ M}^{-1}\text{s}^{-1} \qquad (28)$$

$$OH + O_3 \rightarrow HO_2 + O_2 \qquad 3 \times 10^9 \text{ M}^{-1}\text{s}^{-1} \qquad (29)$$

$$OH + CO_3^{-2} \rightarrow OH^- + CO_3^- \qquad 4.2 \times 10^8 \text{ M}^{-1}\text{s}^{-1} \qquad (30)$$

$$CO_3^- + O_3 \rightarrow \text{products } (CO_2 + O_2^- + O_2) \qquad (31)$$

$$HO_2^- + H_2O = H_2O_2 + OH^- \qquad pK = 11.65 \qquad (32)$$

It should be noted that the factors which can most affect the speciation of ozone itself include changes in pH, the presence of catalytic intermediates such as O_3^-, HO_2^-, O_2^-, OH, H_2O_2, and the presence of high concentrations of dissolved oxygen.

However, recently the decomposition of ozone in basic aqueous solution has been simulated (29-30) in order to better understand the distribution of intermediate species present during the ozone decomposition process. These simulations indicate that O_3^- rather than the OH radical is one of the most important intermediate species and that the rates of the propagation reactions may be comparable in magnitude to the initiation reaction. These findings suggest that further experimental work should be directed toward exploring the nature of the O_3^- radical during both the ozone decomposition and ozone disinfection processes.

ANALYTICAL METHODOLOGY. Because ozone and many of its decomposition products are such powerful oxidizing agents, care must be taken when selecting a specific analytical technique. The analytical reagents used for the determination of ozone are frequently oxidized by ozone, and therefore these same reagents can also be oxidized by many of the oxidation/decomposition products of ozone itself. In addition, many impurities normally found in water are oxidized by ozone to produce oxidizers capable of reacting with the ozone reagents.

The decomposition of ozone involves very reactive and catalytic intermediates which influence the stability of ozone solutions and readily alter the precision and accuracy of various analytical methods for dissolved ozone. These species also influence the role and magnitude of various potential interferents on the ozone decomposition process. From this point of view, it is important to note that many of the older methods for the determination of residual ozone in solution such as the iodometric method are not species specific methods. For example, the iodometric method measures any and all of the oxidizing species present in a decomposing ozone solution such as ozone itself, O_3^-, HO_2^-, O_2^-, OH, HO_2, H_2O_2 (2,31-34). Under normal operating conditions, the total concentration of these highly reactive intermediates is less than 1% of the applied ozone.

In the 1985 Standard Methods description for the determination of residual ozone in water, iodide ion is oxidized to iodine by ozone in an unbuffered potassium iodide solution (31). The pH is then adjusted to 2 with sulfuric acid and the liberated iodine is titrated with sodium thiosulfate to a starch end point. The ozone/iodine stoichiometry for this reaction has been studied extensively (32-34) and found to range from 0.65 to 1.5. Factors affecting the stoichiometry include: pH, buffer composition, buffer concentration, iodide ion concentration, sampling techniques, and reaction time. The pH during the initial ozone/iodide ion reaction and the pH during the iodine determination have been shown to alter the ozone/iodine stoichiometry.

There have been several modifications of the method in an attempt to obtain the necessary one to one stoichiometry. However, when the desired stoichiometry is observed, it appears to be the result of a balance among the various reactions. Ideally the reaction of ozone with potassium iodide is as follows :

$$O_3 + 2I^- + H_2O \ \rightleftharpoons \ I_2 + O_2 + 2OH^- \tag{33}$$

The released iodine is titrated with sodium thiosulfate:

$$I_2 + 2S_2O_3^{2-} \rightleftharpoons 2I^- + S_4O_6^{2-} \qquad (34)$$

Ideally, the iodometric determination of ozone can be reproducible when carried out under very strictly controlled conditions. When used under defined procedural techniques the iodometric method may be useful as an independent check of a UV monitor such as within ± three to five percent. However, the evidence that even microscopic details of the sample bubble passing through the reagent solution can effect the determination makes the iodometric determination of ozone not an ideal candidate as a Standard Method for ozone determination (34).

Additional imprecision in /the measurement of ozone is caused by improper handling of the samples. Since ozone is volatile, any operation which will allow the ozone to diffuse from the sample will produce random error. The use of stripping techniques, although a way to isolate the ozone from potential interference, creates imprecision caused by variations in conditions and enhancement of the decomposition rates.

Ozone rapidly and stoichiometrically decolorizes indigo trisulfonate in acidic solution. The limits of detection are 2 μg/L for an instrumental method and 10 μg/L for a visual field method. This new standard method for ozone is needed because current methods are generally modifications of chlorine residual methods which measure total oxidants. None of these old methods is selective enough for a straight forward determination of ozone in real drinking water or wastewater. The indigo method for the determination of ozone is quantitative, selective and simple.

The method has been tested both for a variety of laboratory conditions as well as for various waterworks which use lakewater, river infiltrate, manganese containing groundwaters, ground waters of extreme hardness and even for its application for biologically pretreated domestic wastewaters. The fact that indigo does not require iodide ion to form intermediates but uses the direct, selective reaction of ozone with the double bond of indigo is the primary reason for the simplicity of the method and lack of interferences.

The Indigo Method was developed by Hoigné and Bader at the Swiss Federal Institute for Water Resource and Water Pollution Control (EAWAG) (35-38). The method is very sensitive, precise, fast and more selective for residual ozone than other methods. Manganese ions, chlorine, hydrogen peroxide and ozone decomposition products, and the products of ozonolysis of organic solutes which interfere with iodometric and similar methods exhibits less interference with the Indigo procedure.

The change in absorbance is followed most accurately by using spectrophotometry although a manual procedure is available (35-38). In all Indigo procedures, the ozone dose must be adjusted to decolorize between 20% and 90% of the Indigo Reagent. For one fixed concentration of the reagent and one predetermined relative ratio of volume of reagent to volume of aqueous ozone, the dynamic range is restricted to a factor of about 4.5. Precision is usually between 1-5% depending on technique, method and equipment used. The stoichiometry is 1:1 at pH 2, and the decrease in absorbance is linear with increasing concentration over a wide range.

The Gas Diffusion Flow Injection Analysis (GD-FIA) technique, can eliminate most, if not all, Indigo Method interferences (in this case chlorine, bromine, and manganese). The GD-FIA Indigo Method exhibits an improvement over the manual method in terms of higher linear range, greater precision among samples, higher sampling frequency, and lower reagent consumption (39).

In GD-FIA the volatile ozone diffuses through a gas permeable hydrophobic membrane into an acceptor stream where the Indigo reaction takes place, and the resulting product is detected. Microporous 0.45 μm pore size Teflon membrane frequently are used because they are chemically inert, an important characteristic when considering the high oxidation potential of ozone. Gas diffusion cells are commercially available with a 75 mm by 2 mm diffusion area.

The GD-FIA procedure eliminates the interference of oxidized forms of manganese, and reduces the interference of chlorine to less than 25% of its value in the flow injection analysis method without gas diffusion. This is equivalent to an error in the measurement of 1 mg/L of dissolved ozone of 0.8% for each mg/L of Cl_2 present (39).

Finally, in the gas phase ozone absorbs ultraviolet radiation with a maximum absorption at 253.7 nm. At 253.7 nm, the generally accepted value for the gas phase absorption coefficient for ozone is 3000 ± 30 M^{-1} cm^{-1} at 273°K and 760 torr (40-42). This value has been approved by the International Ozone Association in Europe.

If the molar absorptivity of 3000 ± 30 M^{-1} cm^{-1} for gaseous ozone is unambiguously accepted as recommended by the IOA European Standardization Committee, UV absorption becomes an absolute method (± 1%) for the determination of gaseous ozone which is not dependent upon calibration or standardization against other analytical methods. Therefore, it can be used for calibration of other analytical methods for ozone.

In conclusion, both ozone and chlorine dioxide appear to be very promising as alternative oxidants and/or disinfectants. Hopefully, we will continue to improve our understanding of the details of the reactions and interactions of these important species.

LITERATURE:

1.   Gordon, G.; Kieffer, R.G.; Rosenblatt, D.H. "The Chemistry of Chlorine Dioxide" in Progress in Inorganic Chemistry, Vol. 15, Lippard, S.J. Editor, (New York, NY: John Wiley and Sons, 1972), pp. 201-286.

2.   Gordon, G.; Cooper, W.J.; Rice, R.G.; Pacey, G.E. Disinfectant Residual Measurement Methods, American Water Works Association - Research Foundation (AWWA-RF ISBN 0-89867-408-5) Denver Colorado, 1987, 815pp

3.   Tomiyasu, H.; Fukutomi, H.; Gordon, G.   "The Kinetics and Mechanism of Ozone Decomposition in Basic Aqueous Solution", Inorg. Chem., 1985, 24, 2962-2966.

4.   Gordon, G.   "The Role of Transition Metal Ions on Oxyhalogen Redox Reactions" J. Pure Appl. Chem.,1989, 61, 873-878.

5.   Aieta, E.M.; Berg, J.D.   "A Review of Chlorine Dioxide in Drinking Water Treatment", J. Am. Water Works Assoc., 1986, 78, 62-72.

6.   Bull, R.J.   "Health Effects of Alternate Disinfectants and Their Reaction Products", J. Am. Water Works Assoc., 1980, 72, 299-303.

7.   Abdel-Rahman, M.S.; Couri, D.; Bull, R.J.   "Kinetics of ClO_2 and Effects of ClO_2, ClO_2-, and ClO_3- in Drinking Water on Blood Glutathione and Demolysis in Rat and Chicken", J. Environ. Pathol. and Toxicol., 1979, 2, 439-449.

8.   Condie, L.W.   "Toxicological Problems Associated with Chlorine Dioxide", J. Am. Water Works Assoc., 1986, 78, 73-78.

9.   Taube, H.; Dodgen, H.   "Applications of Radioactive Chlorine to the Study of the Mechanisms of Reactions Involving Changes in the Oxidation State of Chlorine", J. Am. Chem. Soc., 1949, 71, 3330-3336.

10. Katakis, D.; Gordon, G.   Mechanisms of Inorganic Reactions, John Wiley, (ISBN 0-471-84258-3), 1987, 410 pp.

11. Tang, T-F; Gordon, G.   "Stoichiometry of the Reaction between Chlorite Ion and Hypochlorous Acid at pH 5", Environ. Sci. Technol., 1984, 18, 212-216

12. Emmenegger F.; Gordon, G.   "The Rapid Interaction between Sodium Chlorite and Dissolved Chlorine", Inorg. Chem., 1967, 6, 633-635.

13. Aieta, E.M.; Roberts, P.V.   "Application of Mass-Transfer Theory to the Kinetics of a Fast Gas-Liquid Reaction: Chlorine Hydrolysis", Environ. Sci. Technol., 1986, 20, 44-50.

14. Aieta, E.M.; Roberts, P.V.   "Chlorine Dioxide Chemistry: Generation and Residual Analysis" in Chemistry in Water Reuse, Vol. 1, Cooper, W.J., Editor (Ann Arbor, MI: Ann Arbor Science Publishers, Inc., 1981), pp. 429-452.

15. Themelius, D.; Wood, D.; Gordon, G.   "Determination of Sub-mg/L Levels of Chlorite Ion and Chlorate Ion by Using a Flow Injection System", Anal. Chim. Acta, 1989, (In Press,).

16.   Gordon, G.; Yoshino, K.; Themelis, D.G.; Wood, D.; Pacey, G.E. "Utilization of Kinetic Based Flow Injection Analysis Methods for the Determination of Disinfection Species", Anal. Chim. Acta, 1989, Special Edition on Kinetics in Analytical Chemistry (In Press,).

17.   Pacey, G.E.; Hollowell; D.A. Miller; K.G. Straka; M.R., and Gordon, G.; "Selectivity Enhancement by Flow Injection Analysis", Anal Chim. Acta, 179 (1986) 259-267.

18.   Dodgen, H.; Taube, H. "The Exchange of Chlorine Dioxide with Chlorite Ion and with Chlorine in Other Oxidation States", J. Am. Chem. Soc., 1949, 71, 2501-2504.

19.   Alder, M.G.; Hill, G.R. "Kinetics and Mechanism of Hydroxide Catalyzed Ozone Decomposition in Aqueous Solution", J. Am. Chem. Soc., 1950, 72, 1884-1886.

20.   Sullivan, D.E. "Self-Decomposition and Solubility of Ozone in Aqueous Solutions", Ph.D. Dissertation, Vanderbilt University, December, 1979, 143 pp.

21.   Gurol, M.D.; Singer, P.C. "Kinetics of Ozone Decomposition: A Dynamic Approach", Environ. Sci. Technol., 1982, 16, 377-383.

22.   Roth, J.A.; Sullivan, D.E. "Kinetics of Ozone Decomposition in Water", Ozone Sci. Eng., 1983, 5, 37-49.

23.   Taube, H.; Bray, W.C. "Chain Reactions in Aqueous Solutions Containing Ozone, Hydrogen Peroxide and Acid", J. Am. Chem. Soc., 1940, 62, 3357-3373.

24.   Bühler, R.E.; Staehelin, J.; Hoigné, J. "Ozone Decomposition in Water Studied by Pulse Radiolysis: I. HO_2/O_2^- and HO_3/O_3^- as Intermediates", J. Phys. Chem., 1984, 88, 2560-2564.

25.   Staehelin, J.; Bühler, R.E.; Hoigné, J. "Ozone Decomposition in Water Studied by Pulse Radiolysis: II. OH and HO_4 as Chain Intermediates", J. Phys. Chem., 1984, 88, 5999-6004.

26.   Staehelin, J.; Hoigné, J. "Reaktionsmechanismus und Kinetik des Ozonzerfalls in Wasser in Gegenwart organischer Stoffe" ("Mechanism and Kinetics of Decomposition of Ozone in Water in the Presence of Organic Solutes"), Vom Wasser, 1983, 61, 337-348.

27.   Sehested, K.; Holcman, J.; Bjergbakke, E.; Hart, E.J. "Ultraviolet Spectrum and Decay of the Ozonide Ion Radical, O_3^-, in Strong Alkaline Solution", J. Phys. Chem., 1982, 86, 2066-2069.

28.   Forni, L.; Bahnemann, D.; Hart, E.J. "Mechanism of the Hydroxide Ion Initiated Decomposition of Ozone in Aqueous Solution", J. Phys. Chem., 1982, 86, 255-259.

168

29. Grasso, D.; Weber, W.J.,Jr. "Mathematical Interpretation of Aqueous-Phase Ozone Decomposition Rates", 1989, 115, 541-559.

30. Chelkowska, K.; Grasso, D.; Fabian, I.; Gordon, G. "Mechanistic Comparisons of Residual Ozone Decomposition. New Developments: Ozone in Water and Wastewater Treatment", Proc. Int. Ozone Assoc., 66-76, 1990.

31. Standard Methods for the Examination of Water and Wastewater, 16th Edition, Greenburg, A.E., Trussell, R.R., Clesceri, L.S., Franson, M.A.H., Editors (Washington, DC: American Public Health Assoc., (1985), 426-429.

32. International Ozone Association, Standardisation Committee - Europe "Iodometric Method for the Determination of Ozone in a Process Gas", pp. 001/87 (F), 1987.

33. Gordon, G.; Rakness, K.; Vornehm, D.; Wood, D. "Limitations of the Iodometric Determination of Ozone' J. Am. Water Works Assoc., 1989, 81 (6), 72-76.

34. Wood, D.; Rakness, K.; Vornehm, D.; Gordon, G. "Limitations of the Iodometric Method for the Determination of Ozone" J. Am Water Works Assoc.,1989, 81 (6), 72-76.

35. Bader, H.; Hoigné, J. "Determination of Ozone in Water by the Indigo Method", Water Research 1981, 15, 449-456.

36. Bader, H.; Hoigné, J. "Determination of Ozone in Water by the Indigo Method; A Submitted Standard Method", Ozone: Science and Eng., 1982, 4, 169-176.

37. Bader, H.; Hoigné, J. "Analysis of Ozone in Water and Wastewater by an Indigo Method", presented at 4th Ozone World Congress,sponsored by Intl. Ozone Assoc.; Nov., Houston, TX, 1979.

38. International Ozone Association, Standardisation Committee "Colorimetric Method for the Determination of Residual Ozone in Water (Indigo-Trisulphonate - Method)", pp. 004/89 (F), 1989.

39. Straka, M.R.; Gordon, G.; Pacey, G.E. "Residual Aqueous Ozone Determination by Gas Diffusion Flow Injection Analysis",Anal. Chem., 1985, 57, 1799-1803.

40. International Ozone Association, Standardisation Committee - Europe "Ozone Concentration Measurement in a Process Gas by U.V. Absorption", pp. 002/87 (F), 1987.

41. Hart, E.J.; Sehested, K.; Holcman, J. "Molar Absorptivities of Ultraviolet and Visible Bands of Ozone in Aqueous Solutions", Anal. Chem., 1983, 55, 46-49.

W. J. MASSCHELEIN

Chlorine Dioxide

ABSTRACT

Discovered in 1811 by Sir Humphrey DAVY by reacting potassium chlorate with sulphuric acid, chlorine dioxide is a relatively unstable gas (boiling point : 11°C), which is not to be compressed and liquified without the danger of it exploding. Therefore it must be generated on the site of use.

Chlorine dioxide is an angular free radical with oxidizing properties :

It is a very powerful disinfectant in water; this is due to its properties of high diffusion into hydrophobic lipid layers. The reason for this is that it is a dissolved gas, a free radical or gas-liquid boundary layer and therefore more favourable for contact with vital centres of organisms to be deactivated than are ions.

In 1900, the Spa Waterbaths at Ostend (B) were disinfected by a mixture of potassium chlorate and sulphuric acid. This technology disappeared during World War I and emerged once again during the 50's. Major cities in Europe, such as Düsseldorf, Berlin, Zürich, Toulouse, Brussels, Monaco, use chlorine dioxide in the post-disinfection of their drinking water. A recent inventory of all applications is not available, but we do know that worldwide there exist over one thousand (1,000) sites in operation.

The fact it does not form direct organo-halogen compounds, results in an increase of interest for its use as an alternative to chlorine in water treatment.

W. J. Masschelein, Dr. Sc., Director of Laboratories of the Brussels' Waterboard, Chaussée de Waterloo 764-B-1180 Brussels, Belgium

There are numerous methods to generate chlorine dioxide. The most adequate to the scale of production capacities necessary for water treatment are based at present on sodium chlorite, $NaClO_2$, as a starting product.

A gas phase containing over 10 % vol. ClO_2 is spontaneously explosive and should be avoided or handled accordingly in chlorine dioxide generation.

At the present time, two main reactions are used for the generation of chlorine dioxide for water. The direct reaction of chlorine with dissolved chlorite is faster than the hydrolysis of chlorine :

$$Cl_2 + 2 ClO_2^- = 2 ClO_2 + 2 Cl^-$$

When dissolved in water, chlorine is hydrolysed according to the reaction

$$Cl_2 + H_2 = HOCl + HCl$$

The indirect reaction of hypochlorous acid with ClO_2^- can be summarized according to th following equation :

$$2 H^+ + 2 ClO_2^- + HClO \longrightarrow 2 ClO_2 + H_2O + HCl$$

An alternative process for chlorine dioxide generation is based on direct acidification according to the reaction :

$$5 NaClO_2 + 4 HCl = 4 ClO_2 + 5 NaCl + 2 H_2O$$

Seeing the considerable interest in chlorine dioxide as an oxidant for water treatment, eventually as an alternative to chlorination, extended investigations on the toxicity of residual chlorine dioxide, chlorite and chlorate are to be considered. They can be broadly classified in two groups which are the potential effects of the residual chlorine dioxide as such, and the impact of its inorganic reaction products.

INTRODUCTION

Chlorine dioxide was discovered in 1811 by Sir Humphrey Davy by reacting potassium chlorate with sulphuric acid. Chlorine dioxide is a relatively unstable gas (Boiling Point : 11°C) which cannot be compressed and liquified without explosion danger. Therefore it must be generated at the site of use.

Chlorine dioxide is an oxidant with two consecutive reactions

$$ClO_2 + 1e = ClO_2^- \quad ; \quad E_0^{25°C} = 1.15 \text{ V as gas}$$
$$= 0.95 \text{ V dissolved as liquid}$$

$$ClO_2^- + 4e + 2 H_2O = Cl^- + 4 OH^- \quad ; \quad E_0^{25°C} = 0.78V$$

The oxidation potential of chlorine dioxide in aqueous solution decreases linearly by -0.062 V with each unit increase of pH.

An experimental equation that can be used to evaluate the temperature dependance of the first redox potential of chlorine dioxide is given by

$$E_0 \text{ (volt)} = (-5.367 + 0.0613 \, T - 19.4 \times 10^{-5} \, T^2 + 2 \times 10^{-7} \, T^3)$$

The chlorite ion is a less significant oxidant particularly when reactions with organic compounds are concerned. Reaction products are commented upon later.

Chlorine dioxide gas is potentially explosive when present in air at a partial pressure higher than 100 mbar or 10 % vol. at atmospheric pressure. The explosion may be caused by an electrical discharge, heat or a source of ignition of any kind. To be violent the decomposition needs a temperature above 45°C to 50°C. The reaction is characterized by long induction periods.

Solid sodium chlorite is explosive on heating (or on contact with organic materials (limit of danger : $\pm$ 80°C). The chemical is best not stored in the solid state but in solution e.g. 300-400 g L^{-1}. When spilled the area is to be cleaned by abundant sluiceing with water.

GENERATION OF CHLORINE DIOXIDE

There exist numerous methods to generate chlorine dioxide (10). The most adequate to the scale of production capacities necessary for water treatment are based, at present, on sodium chlorite, NaClO$_2$, as a starting product.

SAFETY CRITERIUM

A gas phase containing more than 10 % vol. ClO$_2$ is spontaneously explosive and should be avoided or handled accordingly in chlorine dioxide generation. The solubility diagrams for chlorine dioxide in water, illustrated in Fig. 1, indicate that at a temperature of 20°C of the process water the maximum "safe concentration" of chlorine dioxide ranges 8-9 g L^{-1}.

The gas density of chlorine dioxide is about 2.4 times that of air.

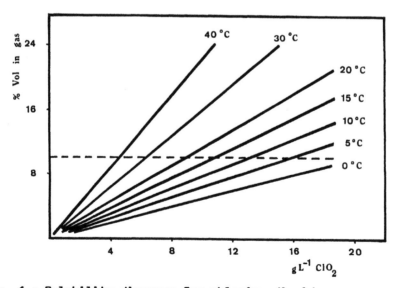

Fig. 1 : Solubility diagrams for chlorine dioxide.

CHLORITE-CHLORINE REACTION FOR GENERATION OF CHLORINE DIOXIDE

Hypochlorous acid and chlorine as Cl_2 both react with sodium chlorite solutions by fast reactions giving chlorine dioxide.

Aieta has shown (1) that the direct reaction of chlorine with dissolved chlorite is faster than the hydrolysis of chlorine.

$$Cl_2 + 2\ ClO_2^- = 2\ ClO_2 + 2\ Cl^-$$

k_2 (zero ionic strength) $= 1.62 \times 10^4$ L M^{-1} s^{-1}

k_2 (at 4 M L^{-1} ionic strength) $= 1.12 \times 10^4$ M^{-1} s^{-1}

Activation energy : 40 ± 5 kJ mol^{-1}

$(k_2 = 1.31 \times 10^4 \times \exp(-40/RT))$

This second order reaction is favourized by chloride ions and assumes (1,2) an intermediate state like

$$2\ |Cl - Cl \underset{O}{\overset{O}{<}} | = 2\ ClO_2 + 2\ Cl^-$$

competition can occur through the secondary reaction

$$|Cl - Cl \underset{O}{\overset{O}{<}} | + H_2O = ClO_3^- + Cl^- + 2\ H^+$$

In usual conditions the half-life time of the direct reaction of chlorine on chlorite ranges 10^{-5} s.

When dissolved in water, chlorine is hydrolysed according to the reaction $Cl_2 + H_2O = HOCl + HCl$

The reaction is of first order, $k_1 \simeq 12.8$ s^{-1} (at 20°C), th.e. a half-life time of 10^{-2} s. The reaction kinetics is only slightly dependent on pH; $k_1 = 12.7$ s^{-1} at pH = 10 and $k_1 = 12.9$ s^{-1} at pH = 3. The activation energy ranges 60 ± 12 kJ mol^{-1}.

The acidity constants of hypochlorous acid are :

t°C	5	10	20	30	35
pK_a	7.754	7.69	7.582	7.497	7.463

This constant determines the respective domains of existence of hypochlorous acid and hypochlorite as indicated in Fig. 2. Except at very low pH-values and very high concentrations of total chlorine "gaseous dissolved Cl_2" does not significantly exist in water, contrary to bromine for example.

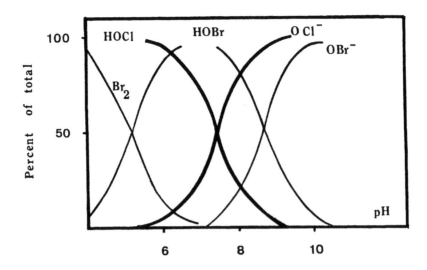

Fig. 2 : HOCl-ClO⁻ equilibrium in water (cfr. dissolved bromine).

No significant formation of chlorine dioxide is obtained on reaction of ClO⁻ with ClO_2^- (3). Marginal formation of chlorate is observed.

The reaction of hypochlorous acid with ClO_2^- can be summarized according to the literature (2) :

$$2\ H^+\ +\ 2\ ClO_2^-\ +\ HClO\ \longrightarrow\ 2\ ClO_2\ +\ H_2O\ +\ HCl$$

The authors also suggest $HClO_2$ as the reactive form of chlorite. At neutral pH however $HClO_2$ does hardly exist :

$$HClO_2\ <==>\ ClO_2^-\ +\ H^+\ ;\ K_a\ =\ 1.1 \times 10^{-2}$$

$$H_2CO_3\ <==>\ HCO_3^-\ +\ H^+\ ;\ K_a\ =\ 4.07 \times 10^{-7}$$

$$HClO\ <==>\ ClO^-\ +\ H^+\ ;\ K_a\ =\ 3.3 \times 10^{-8}$$

These dissociation constants express as indicated in Fig. 3.

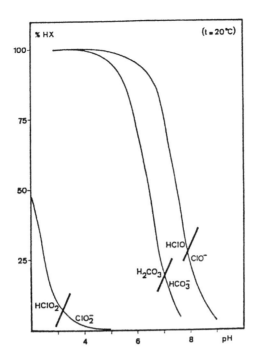

Fig. 3 : Acid-base equilibria relevant to chlorine dioxide generation.

All experimental data converge to consider the formation of chlorine dioxide at neutral pH according to the global stoichiometry as probable:

$HOCl + 2 ClO_2^- = 2 ClO_2 + Cl^- + OH^-$ or also :

$2 NaClO_2 + HOCl + HCl = 2 ClO_2 + 2 NaCl + H_2O$

The question relates also to the final pH-value and the excess chlorine necessary to obtain complete reaction of chlorite.

At too low chlorine concentration the stoichiometry is not obtained, and due to partial reaction, chlorite remains in the reactor effluent. A minimum concentration, or otherwise, an excess of chlorine is necessary (5) to obtain a yield higher than 95 %.

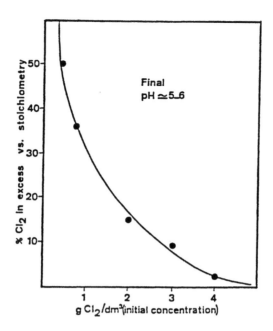

<u>Fig. 4</u> : Excess chlorine necessary for > 95 % conversion of chlorite to chlorine dioxide (5).

In most practical conditions the final pH is about 5 units and may depend slightly on local conditions like the alkalinity of the process water.

Cl_2 (HOCl) + CHLORITE REACTORS.

The technologies of the reactors for generation of chlorine dioxide by oxidation of chlorite with chlorine are generally based on the use of prehydrolized chlorine, th.e. HOCl. They have been described extensively before (11).
Fig. 5 and Fig. 6 illustrate the components of a single flow through reactor capable of produceing 0.5 to 4.5 kgh^{-1}.

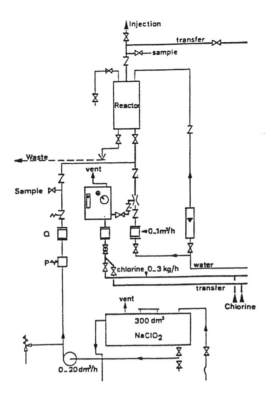

Fig. 5 : Schematic of the ClO_2 generator at the Tailfer plant of the Brussels' Waterboard.

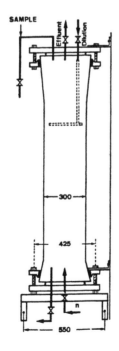

Fig. 6 : Dimensions of the reactor column at the Tailfer plant.

Nominal dosing capacities are :
chlorine : 0 - 3 kg/h   )
dissolution water : 0 - 1 m^3/h   )   1 to 4 g Cl$_2$/dm^3

sodium chlorite 25 % : 0 - 20 L/h
dilution water (optional) : 0 - 5 m^3/h which can also be operated to
                                    rinse the reactor for stand-by.

The flow-through pattern of the reactors are indicated in Fig. 8. The
reactor effluent is directly dosed in the raw water without intermediate
storage. The flow of the 0.75 m^3/s to be treated is sufficiently
constant to adopt this technique. The residual active oxidants are
measured after reaction, the necessary dosing rate is adjusted manually
and the operation controlled by the recording of the different flows of
the chemicals involved in the generation. The analysis required to esta-
blish the ClO$_2^-$-ClO$_3^-$ balance is part of the daily routine work in the
plant operation control. Designed, installed and operated continuously
since 1973, the system has proven to be reliable on long term.

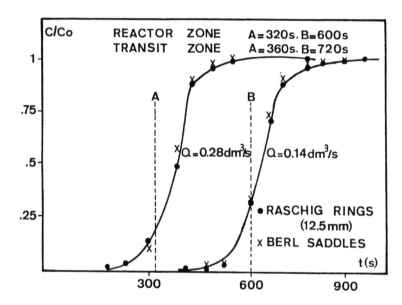

Fig. 7 : Flow-through curves for the Tailfer reactor.

Intermittent generation is sometimes necessary if the demand in chlorine
dioxide is much variable.

In order to adapt the chemical dosing rates as a function of the variations in ClO$_2$ demand, the reactor is sometimes overdimensioned to obtain
continuously correct mixing conditions. The reactor is operated on an
on-off basis to fill an intermediate storage tank. The system is illustrated in Fig. 8.

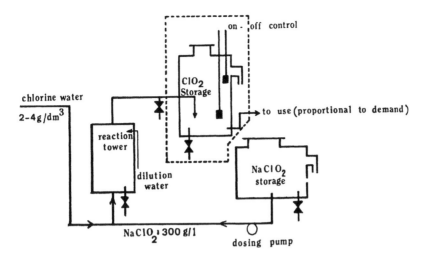

Fig. 8 : Schematic for intermittent generation of chlorine dioxide and proportional dosing.

High performance and fast-acting security systems that control the start-stop sequences are necessary to operate intermittent generating reactors (a response time of within 2 percent of the average nominal residence time in the reactor zone is necessary). The reactor operation time is kept at least at 25 % - 30 % th.e. stand-by periods of less than 75 % of the total cycle of operation. The dilution ratio of the reactor effluent must lower the concentration in chlorine dioxide to under 1 g L^{-1}.

The system as installed at the service reservoirs of the Brussels' area is illustrated in Fig. 9.

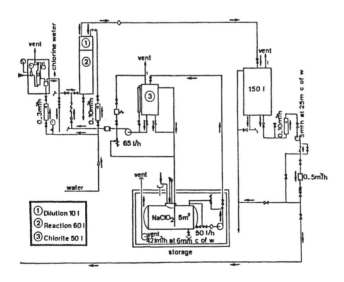

Fig. 9 : Schematic of the discontinuous ClO$_2$ generators.

In normal conditions, the reactor is operated at a maximum 0.75 m^3/h flow of process water which is equal to a theoretical transit time of

300 s. The reactor is normally operated at 2 - 2.5 g Cl_2/dm^3 in order to produce a concentrated chlorine dioxide solution of 4 - 5 g ClO_2/dm^3, which is then diluted 10-fold for storage in the intermediate dosing vessel.

The published data for the direct reaction of molecular chlorine with sodium chlorite to form chlorine dioxide (1,12) always concerned reactions with chlorine in excess to stoichiometry although industrial claims have been made for a production at stoichiometrical proportions with a yield in chlorine dioxide up to 95-98 %.

Excess chlorine in the generation of chlorine dioxide may be detrimental due to the oxidation into chlorate, particularly in acid (pH < 5) medium.

$$H_2O + 2 ClO_2 + HOCl = 2 ClO_3^- + 2 H^+ + HCl$$

$$\frac{d [ClO_3^-]}{dt} = 2 k_2 [HOCl] [ClO_2] \text{ with } k_2 = 0.021 \text{ L M}^{-1} \text{ s}^{-1} (25°C)$$

The number of moles of chlorine dioxide that is reacted is however always slightly higher than the number of moles of hypochlorous acid. A proposed secondary reaction to account therefore is :

$$2 ClO_2 = Cl_2 + 2 O_2$$

ACIDIFICATION OF CHLORITE AS A PROCESS OF GENERATION OF ClO_2.

An alternative process for chlorine dioxide generation is based on direct acidification according to the reaction

$$5 NaClO_2 + 4 HCl = 4 ClO_2 + 5 NaCl + 2 H_2O$$

Hydrochloric acid introducing a common ion in the system is more favourable than other acids. Competition by this reaction giving 4 molecules chlorine dioxide for five moles chlorite engaged is to be avoided in the chlorine-chlorite generation process.

The reaction kinetics of the system $NaClO_2 + HCl$ remains uncompletely investigated but the reaction is much slower than that of chlorine with chlorite and is temperature-dependent (4). Moreover, to operate the process quantitatively, high concentrations are required. The reaction is operated in the conditions of pH in which $HClO_2$ is the dominant form of chlorite (5). Thus, besides the reaction of synthesis

$$5 HClO_2 = 4 ClO_2 + Cl^- + H^+ + 2 H_2O$$

a secondary reaction with lower formation of chlorine dioxide has been reported (8) :

$$4 HClO_2 = 2 ClO_2 + ClO_3^- + Cl^- + H^+ + H_2O$$

Increased concentrations in chloride ion lowers the contribution of the latter reaction whence decreases the concentration of chlorate as a secondary reaction product.

The most important decay reaction for neutral or slightly acid high concentrated solutions of chlorine dioxide is summarized by the global
equation

$$6\ ClO_2\ +\ 3\ H_2O\ =\ 5\ HClO_3\ +\ HCl$$

Therefore also the high concentrated solutions are best diluted to concentrations lower than 1 g L^{-1}.

Two consecutive kinetical phases are to be distinguished in the decomposition of chlorine dioxide solutions (9). At concentrations below 1 gL^{-1}
it lasts more than one hour for the decomposition to start. Pseudo-first
order kinetical constants are approximately $k_0 \sim 5 \times 10^{-6}$ s^{-1} in distilled
water and $k_0 \sim 1.5 \times 10^{-5}$ in saline water (0.27 M, Na_2SO_4). In the
presence of chlorite the pseudo-first order decay constant of chlorine
dioxide has been reported to be increased up to 6.5×10^{-5} s^{-1}.

HCl + NaClO$_2$ reactors

In order to approach the theoretical yield of 4 ClO_2 vs. 5 $NaClO_2$
several conditions must be fulfilled for which general guidelines have
been defined by the C.I.B.E. :

- an excess of HCl is required; example : by working at equal weight of
  HCl and $NaClO_2$;

- on dilution of $NaClO_2$ with process water, the precipitation of $CaCO_3$
  hinders the transfers, whereas unless softened process water is used,
  the concentrated solution of $NaClO_2$ is best injected in pre-diluted
  HCl;

- in laboratory investigation at atmospheric pressure and at open vessel
  reactors, the reaction is best operated in a tubular vessel with a low
  ratio of liquid surface to volume proportion for instance a titration
  burette;

- the best working conditions in full-scale continuous reactors are obtained when carrying out a reacting mixture containing per L $\geqslant$ 45-50 g
  $NaClO_2$ and 50 g HCl, that is 300 % of the stoichiometric quantity of
  the acid. Practical proportions can range 250-300 % of the stoichiometric amounts of HCl;

- instant mixing is essential to obtain a good generation yield. Therefore a built-in mixing baffle in the reacting zone favours the reaction;

- the pH-value of the reactor effluent containing up to 30 g ClO_2 L^{-1} is
  in the range of $\leqslant$ 0.5. If the pH is above 1, the reaction is to slow
  and only partial yields are obtained;

- owing to this high ClO_2 concentration, the whole process is to be operated under vacuum as indicated in Fig. 10. In case of lack of process
  water or vacuum, all security devices must stop the dosing.

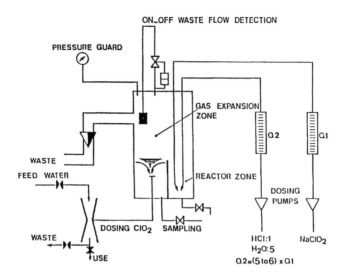

Fig. 10 : Schematic of ClO_2 generation by reaction of $NaClO_2$ with HCl.

A typical reactor is illustrated in Fig. 11. It has a generating capacity of 2.2 kg h^{-1} at nominal capacity and a residence time in the reactor zone of 300 s. The yield is higher than 95 %. Part of the chlorine dioxide is sucked from the reactor as a gas and an other part as a solution.

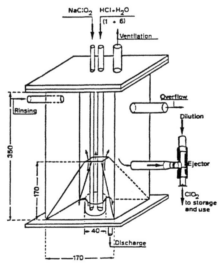

Fig. 11 : Diagram of truncated-pyramid reactor (system : Reservoir of Uccle, Brussels' Waterboard).

The nominal detention time in the reactor zone is a compromise between generation of ClO_2 and its stability in highly concentrated acid solutions. The compromise ends in a reaction time between 200 and 600 s at a process water temperature of 15°C (Fig. 12).

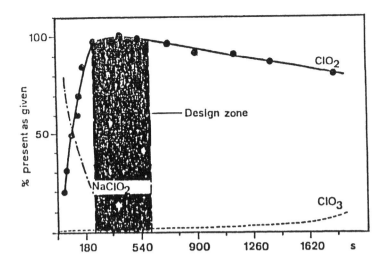

<u>Fig. 12</u> : Reaction time for sodium chlorite (45 g L^{-1}) and hydrochloric
acid (56 g L^{-1}) in a batch reactor (t^o = 15°C).

A more recent study (4) of the reaction conditions of the synthesis of
chlorine dioxide with sodium chlorite and a high excess of hydrochloric
acid, has produced evidence of the temperature effect on the reaction
kinetics. Particularly at lower concentrations, e.g. 10-20 g L^{-1} ClO_2 in
the process water the reaction can be considerably slowed down, e.g. by
a factor of 3, when the temperature is reduced from 20°C to 10°C. In all
instances, at these concentrations, the reaction time was considerably
longer at 10°C than the design zone we have recommended. At first sight
the conclusions were advanced as possibly in contradiction with our data
(5).

However, a more careful examination of the data can give the representa-
tion in Fig. 14. The design zone of about 3 to 6 min. reaction time for
a concentration of 28 to 38 gL^{-1} ClO_2 in the reactor effluent is in fact
confirmed provided the reaction is operated with 280 to 300 % excess of
hydrochloric acid vs. the stoichiometric concentration. In these conditions the temperature of the process water remains of marginal importance at least as far as water temperature in the range of drinking
water are concerned. At lower concentrations of ClO_2, respectively of
ClO_3^-, the time necessary to complete the reaction depends very significantly on the process water temperature. Moreover, the boiling point of
chlorine dioxide is about 11°C.

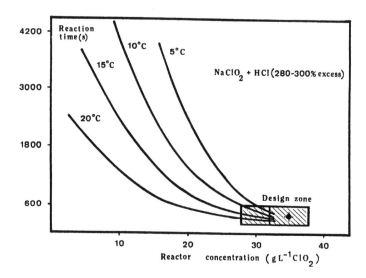

Fig. 13 : Temperature effect on the reaction NaClO$_2$ + HCl.

To meet needs of ClO$_2$ variable in time, discontinuous generation with storage after dilution is feasable as in the case of the NaClO$_2$-Cl$_2$ process.
Constructions that enable to produce variable quantities of chlorine dioxide are f.e. Ben-Ahin reactor of the Brussels' Waterboard (Fig. 14).

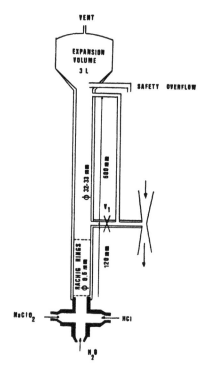

Fig. 14 : Variable rate reactor for generation of chlorine dioxide at Ben-Ahin.

The reactor has two different regimes of operation depending on the position (open-closed) of the valve v_1. The dynamic range of the reactor is of 12 to 150 gh^{-1} ClO_2.

WATER DISINFECTION WITH CHLORINE DIOXIDE

GENERAL

Since the beginning of the 20th century chlorine dioxide is known as a powerful disinfectant of water and was applied in Ostend for Spa waterbaths. During the fifties, it was introduced more generally as a drinking water disinfectant causing less organoleptic problems than chlorine. Important cities in Europe such as Brussels, Zürich, Toulouse, Düsseldorf, Berlin (20), Monaco, Vienna, etc..., and also in the USA (24), Evansville (Ind.), Hamilton (Oh.), Galveston (Tx.), etc. make use of chlorine dioxide. The practical result, as far as disinfection is concerned, is beyond any doubt. However chlorine dioxide must be generated on the site of use. This makes the authorities sometimes hesitant to introduce the use of this less known technology. At present, the problems associated with the formation of THM's and other organo-chlorine compounds illustrate the potential advantages of chlorine dioxide in disinfection of drinking water.

Chlorine dioxide reactions with vital amino-acids are supposed to be one of the dominant processes of bactericidal action. When reacted on amino-acids in excess to chlorine dioxide, a pseudo-first order kinetics on chlorine dioxide is obtained :

ClO_2 - histidine : $T_{1/2}$ = 12 s at pH = 9, and, 1050 s at pH = 7

ClO_2 - proline : $T_{1/2}$ = 60 s at pH = 9 and very slow reaction
at pH = 7

The amino-acids that react significantly with chlorine dioxide are : hydroxyproline, proline, histidine, cysteine, cystine, tyrosine and tryptophane (14).

The pseudo-first order decay constants are then correlated to the concentration of the disinfectant according to Watson's law : $C^n \times t = k$

For E-coli the concentration factor n is 1 for chlorine while it ranges about 2 for chlorine dioxide. For Pseudomonas fluorescens (putida) we observed a concentration factor of about 2.5 with chlorine dioxide. For N. gruberi the n value has been established as 1 but the contact time would intervene with a power factor of 1.5 to 2 (16) or, in other words, by setting the contact time with power one, n is in the order of 0.55. We have measured the same order of value for actinomycetaceae in water.

The activity of a bactericidal agent in water can be expressed in terms of dose in $L^{-1} \times mg \times s$ to obtain a given decay, e.g. 99 % kill is expressed as 2 D_{10}.

Typical 2 D_{10}-values in $L^{-1} \times mg \times s$	References	Unit D_{10} or CT-value ($L^{-1}mg$ min) (**)
E.coli : **16** pH = 7, t° = 5°C	(24)	0.14
9 pH = 7, t° = 15°C	(20,21)	0.075
8 pH = 7, t° = 25°C	(24)	0.07
Ps. fluorescens **5-10** (pH = 7-8, t° = 22°C)		0.06 (*)
Naegleria gruberi : **360-600** (pH=7, t°=20°C)	(16)	5
Poliovirus : **100** (pH = 7, t° = 15°C)	(21)	0.83
240 (pH = 7, t° = 5°C)	(19,22)	2
750 (pH = 10, t° = 20°C)	(18)	6
1200 (pH = 7, t° = 20°C)(wild strain)	(25)	10
Coxsackie-virus A9 : **22** (pH = 7, t° = 15°C)		0.2
Rotavirus SA11 : **30** (pH = 6, t° = 5°C, purified virus)	(17)	0.25
1000 (pH = 9.5, t° = 5°C, cell associated virus)	(17)	8.3
Bacteriophage f2 : **16** (pH = 7.1, t° = 3°C)	(14)	0.13
Legionella pneumophila **< 5** (pH = 7-8, t° = 22°C)(*)		0.04
Giardia/Lamblia **8400** (pH = 7-8, t° = 15°C)	(24)	70
Clostridium perfringens **4000** (pH = 7, t° = 20°C) (important lethal lag)	(25)	33
Actinomyces (pH = 7, t° = 20°C)(wild strain) (*)		50

(*) Tentative values under investigation
(**) CT-values are computed on the basis of the assumption that the concentration factor n is equal to unit.

Experimental difficulties are associated with obtaining good data under laboratory conditions and with the control of small residual concentrations of chlorine dioxide like those that have been proven efficient in practice, that is equal to residual concentrations under 0.1 mg L^{-1} (15).

Most of the authors did not rely on specific and reliable methods for the analytical control of residual chlorine dioxide but consider the concentration as introduced into the sample water. A probable overestimate of the concentrations is existing in the literature which means an underestimate of the bactericidal action.

EFFECTS OF RESIDUAL CHLORITE

The stable secondary products of chlorine dioxide, like chloride and chlorate, are of no direct value in disinfection.

Chlorite has a weak bactericidal action on some germs possibly present in water like Streptococcus faecalis but the action is too slow to be an operational method of disinfection (26).

The action on Pseudomonas putida is more significant but still slow and other species like Ps. deva are more resistant but significant aftergrowth is damped by chlorite (27). The action on actinomyces and yeasts like Rhodontorula is neglectible.

As a conclusion : residual chlorite can contribute to bactericidal action but is not sufficient. It prevents however significantly after-growth in distribution systems and contributes in this way to the water quality.

Investigations are underway to check how far the "toxic properties" attributed to chlorite can be used to control the development aquatic organisms which have to rely on respiration mechanisms in which inter-vene substances similar to hemobgobin. Fig. 20 illustrates first <u>tenta-tive</u> data for adult Asellus aquaticus.

Chlorite will undoubtly prevent the growth and the development of such organisms in distribution systems. Further investigations are underway (28); age and stress conditions e.g. light may influence the quantita-tive results.

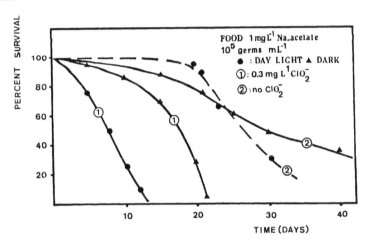

<u>Fig. 15</u> : A sample set of data for the survival of Asellus aquaticus in presence of chlorite ion. (28)

DEVELOPMENT OF BY-PRODUCTS BY REACTIONS OF CHLORINE DIOXIDE IN WATER

GENERAL

The reactions of chlorine dioxide with organic substances cannot be con-sidered separately from the global context of oxidation. In particular, the ammonium and bromide content of the water are of utmost importance to block secondary chlorination or bromination occuring as side reac-tions of the chlorine dioxide treatment.
Chlorine dioxide does not have a direct significant reaction with ammo-nium ions. In the dark bromide ion is also not oxidized with significant velocity by chlorine dioxide. This is in contrast with the reaction pathways with hypochlorous acid and ozone which form hypobromous acid. Photosensitized oxidation of bromide is possible with chlorine dioxide.

REACTION PRODUCTS OF ORGANIC COMPOUNDS

The reaction products of chlorine dioxide with organic compounds in water have been reviewed previously (30,31,32).

SUMMARIZING

* Practically no THM's are formed on the final treatment of drinking water with pure ClO_2. In the case of the raw water of the River Meuse in Tailfer, treated exhaustively with either chlorine or chlorine dioxide the level of TOX's formed with chlorine dioxide is less than 4 % of that formed with chlorine (29). This formation can be minimized, e.g. if ammonium is present in the treated water source.

* Phenols and phenolic compounds undergo ring opening reactions. Di- and tricarboxylic acids and glyoxalic acid are stable end-products. At higher concentration of phenols, of which the 4-position is not substituted, transcient formation of quinones is observed. If the ratio of chlorine dioxide is unsufficient secondary chlorination can eventually occur in the case where no ammonium is present. The reaction of the phenate ions is much faster than that of the corresponding phenols (about 10^5 to 10^6 times).

* Chlorine dioxide reacts with humic acids. The identified reaction products are carboxylic acids, aldehydes, glyoxal and if the reaction is forced by overdosing of ClO_2, there is also the formation of monochloracetic acid besides traces of monochlorosuccinic acid, di- and trichloracetic acid. The decarboxylation of fulvic acid represents about 30 % of the original carbon of the acid oxidized by ClO_2. Fulvic acid exhausts a ClO_2 demand at a ratio of 0.3 moles ClO_2 per mole carbon. Methylesters are also secondary reaction products.
  Model hydroxylated phenyl-carboxyacid are decarboxylated in the first phase of reaction giving quinones in a following step if only a moderate excess of ClO_2 is applied.

* Aliphatic hydrocarbons do not react significantly with chlorine dioxide in practical conditions of water treatment. Aromatic hydrocarbons such as benzo(a)pyrene, anthracene and benzo(a)anthracene react very quickly with chlorine dioxide. Quinones are the major initial reaction products. Pyrene, benzo(e)pyrene, naphthalene and fluorenthene react only very slowly.
  Some olefins are sufficiently activated to react with chlorine dioxide e.g. in styrenes, methyl styrenes and stilbenes. The general reaction scheme then formulated is

$$2 >\!C = C\!<\; +\; 2\ ClO_2\; +\; 2\ H_2O\; \xrightarrow{H^+}\; >\!\overset{\overset{\displaystyle OH}{|}}{C} - \overset{\overset{\displaystyle OH}{|}}{C}\!<\; +\; >\!\overset{\overset{\displaystyle OH}{|}}{C} - \overset{\overset{\displaystyle Cl}{|}}{C}\!<\; +\; H^+\ ClO_3^-$$

which however implicates an acid disproportionation of

$$2\ HClO_2\; \longrightarrow\; HClO\; +\; H^+\; +\; ClO_3^-$$

The half reaction value $t_{1/2}$ depends on the compound and varies between a few minutes to 50 hours (33).

* Amino acids without specific reactive groups do not react with chlorine dioxide under water treatment conditions. This is the case e.g. for glycine, alanine, phenylalanine, serine and leucine. The aromatic part of the molecule or the sulphide groups of reactive amino acids such as tyrosine, tryptophane, histidine, cystine and methionine react with ClO_2. Sulphur-containing amino acids e.g. cystine react with

188

chlorine dioxide, the parent sulphonic acid being the final product with intermediate formation of bisulphoxide. Methionine is converted to the sulphone. The reactions of such amino acids with ClO_2 play an important part in the virucidal action of chlorine dioxide (34).

* Primary amines do not react significantly with chlorine dioxide. Tertiary amines react on the α-carbon site next of the nitrogen followed by a cleavage of the C-N link. In the case of triethylamine, the over-all reaction is :

$$H_2O + 2\ ClO_2 + (C_2H_5)_3N \longrightarrow CH_3CHO + (C_2H_3)_2NH + 2\ H^+ + 2\ ClO_2^-$$

The second order rate constant is approximately $2\times10^5\ LM^{-1}\ s^{-1}$ (35). Investigations on the potential mutagenicity of reaction products of chlorine dioxide with organic products occurring naturally in water like humic acids have proven this effect always being absent.

INORGANIC REACTION PRODUCTS

Chlorite is undoubtly one of the principal reaction products. In the practice of treatment, between 40 to 60 % of the chlorine dioxide reacted is found in the form of chlorite. In model studies this proportion ranges 50 to 100 % according to the experimental conditions, e.g. excess, pH, presence of ammonia, etc.
Part of the ClO_2 is transformed into chlorate either through disproportionation or by photodecomposition through artificial irradiation or by sunlight. Laboratory experiments must be analyzed carefully taking into account the possible disproportionation.

To avoid secondary effects of HClO the presence of ammonia is recommended.

The chlorite ion is a less significant oxidant particularly when reaction with organic compounds or disinfection is concerned. However some reactions of mineral products are of particular interest in the context of water treatment e.g. oxidation of manganese salts into manganese dioxide with formation of chloride. The oxidation does not proceed further than the Mn^{4+} stage, whence eventual overdosing of ClO_2 cannot hinder the demanganization contrary to the use of other oxidants.

$$2\ ClO_2 + Mn^{++} + 4\ OH^- \longrightarrow MnO_2 + 2\ ClO_2^- + 2\ H_2O$$
$$Also \quad 2\ ClO_2 + 2\ S^= \longrightarrow 2\ Cl^- + SO_4^= + S\downarrow$$

TOXICITY OF CHLORINE DIOXIDE AND RELATED OXYCHLORINE COMPOUNDS

Seeing the considerable interest in chlorine dioxide as an oxidant for water treatment, eventually as an alternative to chlorination, extended investigations on the toxicity of residual chlorine dioxide, chlorite and chlorate are to be considered. They can be broadly classified in two groups which are the potential effects of the residual chlorine dioxide as such, and the impact of its inorganic reaction products.

SUMMARY OF TOXICITY DATA ON CHLORINE DIOXIDE AND RELATED IONS IN WATER TREATMENT

(doses and/or concentrations as indicated; literature references within brackets)

EFFECT & PARAMETER UNIT	ClO_2	ClO_2^-	ClO_3^-	REMARKS
Lowest effect level (water mg L^{-1})	12 (38,39,40,42)	1.2 (38)	1.2 (37)	Acute effect : 1000-1500 for ClO_3^- (49)
NOAEL mg kg^{-1} d^{-1}	1, (rats)	0.7-1, (rats)		
LD$_{50}$ mg kg^{-1}	140 (rats)	140-200 (rats)	200 (humans)	
Chronic toxicity, mg d^{-1} / mg L^{-1}	none at 5 (40) { none at 10(50) positive at 30 (50) }	none at 5 (40)	none at 5 (40) / --	5 as sum of the three parameters (humans)
Methemoglobinaemia through water, mg L^{-1}	100 (41,48)	50 (none at 10)(45)	10-100 (45) (rats,chicken)	All reversible up to 1000
Hemolytic effects through water, mg L^{-1}	none 2, positive (45) { 15 pigeon 100 chicken (41) 30 monkey }	100 (rats) (41) none up to 70(41) stress at 50(45)	6-300 (monkey)(45)	Distinction between methemoglobinaemia and hemolysis is arguable.
Glutathion loss, mg L^{-1}	50 (41)	50 (41)	10-100 (41)	
Effect on kidney's, mg L^{-1}	--	none at 100 (mice)(44)	--	Tests in vitro, no effects observed on humans (42)
Embryotoxicity, mg kg^{-1} d^{-1}	--	none at 4, (43) positive at 14 (rats)	--	
Mortality pregnant rats, mg kg^{-1} d^{-1}	--	none at 20,(43) positive at 100	--	
Teratogenic effects, mg kg^{-1} d^{-1}	NOAEL : 1 (rats) 10 (rats)(41,43)	NOAEL : 1 (rats) 10 (rats) (43)	NOAEL : 1 (rats) 10 (rats)(43)	
Carcinogenic effects, mg L^{-1}	--	--	none at 10(45)	
Increased intestinal turnover, mg L^{-1}	>10	>10	not observed	
Body weight effect, mg kg^{-1} d^{-1}	5 (rats)(36)	2 rats (43)	--	
Threshold waterbodies, mg L^{-1}	0.25 (47)	--	10 (46,51)	For humans
Organoleptic threshold water, mg L^{-1}	0.4 (36)	0.3 (47)	5 (46,52)	

The maximum residual concentration of chlorine dioxide allowed in Belgium is 0.25 mg L^{-1}, in the Federal Republic of Germany is 0.2 mg L^{-1}, and in Switzerland is 0.15 mg L^{-1}.

Another approach is that of the Suggested No-Adverse-Response Levels (SNARL's) as published in 1987 by the National Academy of Sciences in 1987 (USA). The calculated values are based on the assumption that 20 % of the daily intake comes from ingestion of drinking water and a safety factor of 10 was taken into consideration. On this basis, figures for the maximum admissible concentrations were advanced as 0.21 mg L^{-1} for ClO_2 and 0.024 mg L^{-1} "for chlorite and chlorate" ions. However, the experimental evidence was poor and not accepted. Indeed known existing concentrations which have proven not to have any effect are accepted as such without considering that any higher concentration could possibly have no effect either.

It is also remarkable that chlorite and chlorate ions are considered indistinctly, although the toxicity levels are probably very different. A lack of accurate knowledge of the real concentrations present is probably the reason for these assumptions.

EPA (US-Federal Register, 52 (212), 42177-42222 (1987)) did not accept the approach of the US-National Academy seeing that none of the compounds considered is expected to be part of the average daily intake otherwise than through drinking water. On the other hand a higher safety factor than ten could be taken in the future.

Assuming that nearly 100 % of exposure comes from drinking water the SNARL is estimated at 1 mg L^{-1} including chlorine dioxide and its oxidation-reduction products chlorite and chlorate.

The association of "chlorate toxicity" with that of chlorite is certainly questionable and the analytical evidence for chlorate is often poor. Whatever the conclusions are, it appears that at present much more is known on the toxicity of chlorine dioxide and related compounds in water than is for chlorine in water.

CONCLUSION

The disinfecting capacity of chlorine dioxide is well recognized since three quarters of a century and is currently used for post-disinfection or safety treatment of drinking water of several major European cities.

All tests made thusfar have indicated the absence of any relevant toxicity of reaction or by-products of the treatment when performed adequately.

Besides the reaction of inorganic compounds like iron, manganese and sulphide, organic phenolic type compounds are modified to give no more organoleptic hindering compounds. Disinfection capacity is far higher than that of chlorine.

(19) O. SPROUL, Y.S.R. CHEN, J.P. ENGEL & A.J. RUBIN, Env.Techn.Letters, 4, 335 (1983).

(20) C.H. RUPP, Ges.Ingenieur, 104, 278 (1983).

(21) P.V. SCARPINO, S. CRONIER, M.L. ZINK & F.A.O. BRIGANO, Proceed. 5th Annual AWWA Conf. on Water Quality in the Distribution System, 2 B-3, p. 1 (1977).

(22) G.R. TAYLOR & M. BUTLER, Journ. of Hyg., 89, 321 (1983).

(23) J.E. PEETERS, E.A. MAZAS, W.J. MASSCHELEIN, I.V. MARTINEZ DE MATURANA & E. DEBACKER, Appl. Env. Microbiology, 55, 1519 (1989).

(24) R.C. RICE, EPA-Workshop, March 1988.

(25) R.S. FUJIOKA, M.A. DOW & B.S. YONEYAMA, Wat.Sci.Techn., 18, 125 (1986).

(26) W.J. MASSCHELEIN, Water S.A., 6, 117 (1980).

(27) W.J. MASSCHELEIN, G. FRANSOLET & E. DEBACKER, Eau du Québec, 14, 41 (1981).

(28) N. HANSEN, G. FRANSOLET & W.J. MASSCHELEIN, Laboratories Brussels' Waterboard, Journ. franç. d'Hydrologie appliquée (in press).

(29) R. SAVOIR, L. ROMNEE & W.J. MASSCHELEIN, Aqua, 2, 114 (1987).

(30) W.J. MASSCHELEIN, Chlorine dioxide, Ed. Ann Arbor Science, Michigan (1979).

(31) Ch. Rav. ACHA, Water Res., 18, 1329 (1984).

(32) M. DORE, Chimie des Oxydants & Traitement des Eaux, Ed. Technique et Documentation, Paris (1989).

(33) Ch. Rav. ACHA, E. CHOSHEN & S. SAREL, Helv. Chim. Acta, 69, 1728 (1986).

(34) C.I. NOSS, F.S. HAUCHMAN & V.P. OLIVIERI, Wat. Res., 20, 351 (1986).

(35) D.H. ROSENBLATT, A.J. HAYES, B.L. HARRISON, R.A. STREATY & K.A. MOORE, J. Org. Chem., 28, 2790 (1963).

(36) S.A. FRIDLYAND & G.Z. KAGAN, Hygiene and Sanitation, 36, 190 (1971).

(37) G.J. TUSCHEWITZKI & J.U. HOHN, 38e Conférence Internationale du Cebedeau, Brussels 1985, p. 261.

(38) J.R. LUBBERS & J.R. BIANCHINE, Journal Env. Pathol. & Toxicol., 5, 215 (1984).

(39) J.R. LUBBERS, S. CHAUHAN, J.K. MILLER & J.R. BIANCHINE, Journal of Env. Pathol. & Toxicol., 5, 229 (1984).

(40) J.R. LUBBERS, J.R. BIANCHINE & R.J. BULL, Chap. N° 95, p. 1335, in Water Chlorination, vol. 4, book 2, Ed. Ann Arbor Science (1983).

(41) D. COURI, M.S. ABDEL-RAHMEN & R.J. BULL, Env. Health Perspectives, 46, 57 (1982).

(42) J.R. LUBBERS, S. CHAUAN & J.R. BIANCHINE, Env. Health Perspectives, 46, 57 (1982).

(43) D. COURI, C.H. MILLER, R.J. BULL, J.M. DELPHIA & E.M. AMMAR, Env. Health Perspectives, 46, 25 (1982).

(44) G.S. MOORE & E.J. CALABRESE, Env. Health Perspectives, 46, 31 (1982).

(45) L.W. CONDIE, Journ. AWWA, 78, 73 (1986).

(46) I.I. AVEZBAKIEV & N.M. DEMIDENKO, Gig. i Sanitarya, 5, 11 (1979).

(47) Brussels' Waterworks, practical experience.

(48) G.S. MOORE, E.J. CALABRESE, S.R. DINARDI & R.W. TUTHILL, Medical Hypotheses, 4, 481 (1978).

(49) J. O'GRADY & E. JARECSNI, British Journal of Clinical Practice, 25, 38 (1971).

(50) J. MUSIL, Z. KNOTEK, J. CHALUPA & P. SCHMIDT, Technology of Water, 8, 327 (1964).

(51) O. PRAVDA, Hydrobiologia, 42, 97 (1973).

(52) D. STOFEN, Stadtehygiene, 24, 109 (1973).

CYNTHIA J. LANGLOIS
EDWARD J. CALABRESE

The Interactive Effect of Chlorite, Copper and Nitrite on Methemoglobin Formation in Red Blood Cells of Dorset Sheep

ABSTRACT

Simultaneous exposure to chemicals which can oxidize the hemoglobin of the red blood cell to methemoglobin is common. Although the effects of some of these agents have been documented individually, little research considers the interactive effects. In vitro experimentation on treated blood of female Dorset sheep assessed the interactive capacity of chlorite, copper and nitrite to affect methemoglobin formation. All combinations of doses which produced 2.5, 5, 10 percent methemoglobin and controls were tested in all possible combinations (a total of 80). This therefore, included data on each chemical alone, each two-way combination and the three-way combination. The response is largely additive (the sum of the individual effects) except for one of the two-way interactions, chlorite/nitrite (p < .01) which showed antagonism. Chlorite may oxidize nitrite which could explain the less-than-additive response. Overall, the result of combining these agents on methemoglobin was additive.

INTRODUCTION

The primary function of the red blood cell is to maintain aerobic metabolism by transporting oxygen to, and carbon dioxide from, body tissues. The iron atom of the heme molecule must be in the ferrous (+2) oxidation state in order for this to occur. Certain chemical processes within the cell, however, oxidize the ferrous iron to the ferric (+3) state. This molecule is known as methemoglobin and is unable to bind oxygen, or carbon

Cynthia J. Langlois and Edward J. Calabrese, Environmental Health Sciences Program, School of Public Health, University of Massachusetts, Amherst, Massachusetts, U.S.A.

dioxide and is not dissociable. Fortunately, there are enzyme systems in the cell which reduce methemoglobin back to hemoglobin. In fact, normally only 1 - 2% of hemoglobin is in the methemoglobin state [1].

Methemoglobinemia can be due to congenital deficiencies of the reducing systems, abnormalities of the hemoglobin itself, or exposure to oxidant chemicals. It is possible that people with deficiencies related to methemoglobin (i.e. Glucose-6-Phosphate Dehydrogenase) are more susceptible to methemoglobin formation by oxidant chemicals [2]. This study considered chemical exposure by testing three known methemoglobin-producing agents (chlorite, copper and nitrite). Although the effects of these chemicals has been documented individually [1], little or no research has considered the interactive effects. The possibility must be considered here since all of these chemicals can be found in drinking water and other sources as well.

Chlorine dioxide is being considered as an alternative water disinfection method by the Environmental Protection Agency. The by-products of chlorine dioxide are chlorite ($ClO2-$), and to a lesser degree chlorate ($ClO3-$). Both are known methemoglobin producers, although chlorite is more potent. Approximately 5% of treatment facilities serving major utilities chose chlorine dioxide as the disinfection method [3].

Copper is important because exposure to it is widespread. Copper is commonly found in drinking water since it can leach from the copper pipes that transport it. This is more problematic in acidic water. Copper is also found in foods (vegetables), air and soil.

The connection between nitrates and methemoglobin is well known. Because of the use of nitrogen-based fertilizer very high levels of nitrates and nitrites have been, and continue to be, found in well water (especially in the Mid-West) [4]. It can be found in some vegetables and is also used as a meat preservative, so it is found in hot dogs, bacon, cold cuts, etc. None of these agents is exotic; they are ubiquitous and exposure to them is common.

METHODS AND MATERIALS

For this in vitro study, whole blood was drawn from six non-pregnant female Dorset sheep. Sheep were chosen as a model because unlike rodents, their methemoglobin-reducing capabilities are similiar to humans [5]. Also, sheep are a model for G-6-PD deficiency.

Blood was incubated with the chemical(s) for one hour in a 37 C bath. Four levels of each chemical were tested in all possible combinations. The levels are ones which produced 0, 2.5, 5 and 10% methemoglobin. The concentrations are 0.125 millimolar (mM), 0.25 mM, and 0.5 mM for nitrite and copper; and 0.5 mM, 1 mM and 2 mM for chlorite. Methemoglobin levels were detected spectrophotometrically using the method of Brown in Hematology: Principals and Procedures [6].

The data were analyzed with the Analysis of Variance statistical test via the BMDP statistical software package.

Although a variety of methods and terminologies exist for assessing interactions, the following definitions are used in this experiment. When the effect of a combination of chemicals is the same as the algebraic sum of all the individual chemicals, the response is known as additive. A synergistic response is one that is significantly greater than additive; while antagonism refers to a significantly less than additive response. The shape of the dose-response curves also suggests the type of interaction. It generally assumed that the effect of a mixture will be additive, although it is certainly not always the case.

RESULTS

See Table 1 for means and standard deviations; Table 2 for Observed vs Expected values. As previously mentioned, this study assessed the effect of each chemical alone, each combination of two chemicals and the three chemical combination. Both nitrite and chlorite alone elicited the percent methemoglobin that was expected based on range finding studies. Copper, however, gave consistantly lower than expected values. Statistically, each of the chemicals alone significantly caused the production of methemoglobin.

The two-way interactions between copper/chlorite and copper/nitrite were strictly additive which was expected. Figures 1 and 2 show the parallelism of the dose-response curves which further suggests additivity. One interesting finding is that the other two-way interaction, chlorite/nitrite, was not additive. The observed values were significantly lower than the expected ones. Also, the dose-response curve shows a clear deviation from parallelism and additivity (Fig. 3). This antagonism was statistically significant as well.

When all three chemicals were combined, the result was slightly less than additive at some points, but this was not statistically significant. The slight deviation from

TABLE 1: RESULTS OF STUDY — IN MEAN PERCENT METHEMOGLOBIN ± STANDARD DEVIATION

Nitrite Levels	0 mM Chlorite				0.5 mM Chlorite				1 mM Chlorite				2 mM Chlorite			
	0 mM	0.125 mM	0.25 mM	0.5 mM	0 mM	0.125 mM	0.25 mM	0.5 mM	0 mM	0.125 mM	0.25 mM	0.5 mM	0 mM	0.125 mM	0.25 mM	0.5 mM
0 mM Copper	1.66 ± 0.44	3.34 ± 0.77	4.95 ± 0.50	11.37 ± 1.03	2.32 ± 0.82	4.60 ± 1.08	6.43 ± 1.26	11.41 ± 1.04	3.94 ± 0.85	6.02 ± 1.21	8.15 ± 1.04	12.56 ± 1.33	7.67 ± 2.14	8.39 ± 2.49	9.19 ± 1.79	13.73 ± 1.33
0.125 mM Copper	2.77 ± 0.92	4.68 ± 1.05	6.83 ± 1.40	12.17 ± 1.14	3.28 ± 0.92	5.27 ± 1.47	7.70 ± 0.98	12.03 ± 0.71	4.49 ± 1.36	6.31 ± 1.85	9.34 ± 1.29	12.42 ± 1.34	7.43 ± 2.35	9.36 ± 2.74	10.99 ± 1.49	14.43 ± 1.81
0.25 mM Copper	4.02 ± 1.61	6.06 ± 1.18	7.80 ± 0.91	13.63 ± 2.43	4.72 ± 0.99	6.67 ± 1.87	9.11 ± 1.62	13.40 ± 1.26	6.45 ± 1.49	8.21 ± 2.82	8.73 ± 1.77	14.01 ± 2.07	8.80 ± 1.99	8.75 ± 2.38	10.63 ± 1.30	15.65 ± 3.10
0.5 mM Copper	5.15 ± 3.02	7.29 ± 2.14	9.69 ± 1.84	16.24 ± 2.39	6.37 ± 2.72	8.48 ± 2.63	10.79 ± 1.09	15.56 ± 2.41	8.69 ± 1.86	10.58 ± 4.18	10.78 ± 1.64	17.05 ± 3.04	10.58 ± 2.72	12.57 ± 3.05	12.24 ± 2.18	18.93 ± 3.76
0.5 mM Na Acetate	1.35 ± 0.73	3.49 ± 0.71	5.51 ± 1.42	11.15 ± 1.29	2.82 ± 0.79	3.97 ± 1.33	7.03 ± 2.40	10.67 ± 0.67	4.49 ± 0.88	5.56 ± 1.42	8.02 ± 2.80	10.79 ± 2.40	6.56 ± 1.24	7.88 ± 2.16	10.31 ± 1.74	14.82 ± 3.38

TABLE 2: OBSERVED VERSUS EXPECTED VALUES

Nitrite Levels	0 mM Chlorite				0.5 mM Chlorite				1 mM Chlorite				2 mM Chlorite			
	0 mM	0.125 mM	0.25 mM	0.5 mM	0 mM	0.125 mM	0.25 mM	0.5 mM	0 mM	0.125 mM	0.25 mM	0.5 mM	0 mM	0.125 mM	0.25 mM	0.5 mM
0 mM Copper	1.66	3.34	4.95	11.37	2.32	4.60	6.43	11.41	3.94	6.02	8.15	12.56	7.67	8.39	9.19	13.73
						4.00	*5.61*	*12.03*		*5.62*	*7.23*	*13.65*		*9.35*	*10.96*	*17.38*
0.125 mM Copper	2.77	4.68	6.83	12.17	3.28	5.27	7.70	12.03	4.49	6.31	9.34	12.42	7.43	9.36	10.99	14.43
		4.45	*6.06*	*12.48*	*3.43*	*6.77*	*8.38*	*14.80*	*5.05*	*8.39*	*10.00*	*16.42*	*8.78*	*12.12*	*13.73*	*20.15*
0.25 mM Copper	4.02	6.06	7.80	13.63	4.72	6.67	9.11	13.40	6.45	8.21	8.73	14.01	8.80	8.75	10.63	15.65
		5.70	*7.31*	*13.73*	*4.68*	*8.02*	*9.63*	*16.05*	*6.30*	*9.24*	*11.25*	*17.67*	*10.03*	*13.37*	*14.98*	*21.40*
0.5 mM Copper	5.15	7.29	9.69	16.24	6.37	8.48	10.79	15.56	8.69	10.58	10.78	17.05	10.58	12.57	12.24	18.93
		6.83	*8.44*	*14.86*	*5.81*	*9.15*	*10.76*	*17.18*	*7.43*	*10.77*	*12.28*	*18.80*	*11.16*	*14.50*	*16.11*	*22.53*
0.5 mM Na Acetate	1.35	3.49	5.51	11.15	2.82	3.97	7.03	10.67	4.49	5.56	8.02	10.79	6.56	7.88	10.31	14.82
		3.03	*4.74*	*11.06*	*2.01*	*5.35*	*6.96*	*13.38*	*3.63*	*6.97*	*8.58*	*15.00*	*7.36*	*10.70*	*12.31*	*18.73*

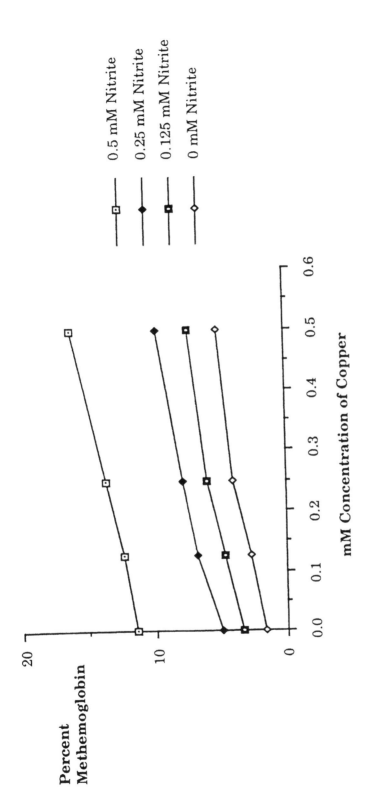

FIGURE 1: COMBINATION COPPER AND NITRITE CONCENTRATIONS
VERSUS PERCENT METHEMOGLOBIN

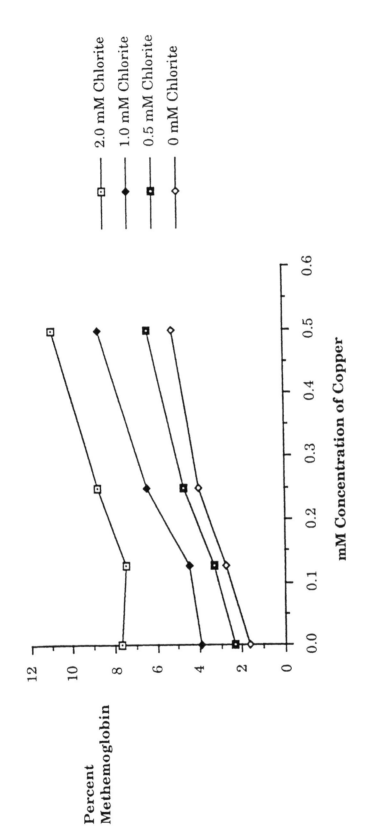

FIGURE 2: COMBINATION CHLORITE AND COPPER CONCENTRATIONS
VERSUS PERCENT METHEMOGLOBIN

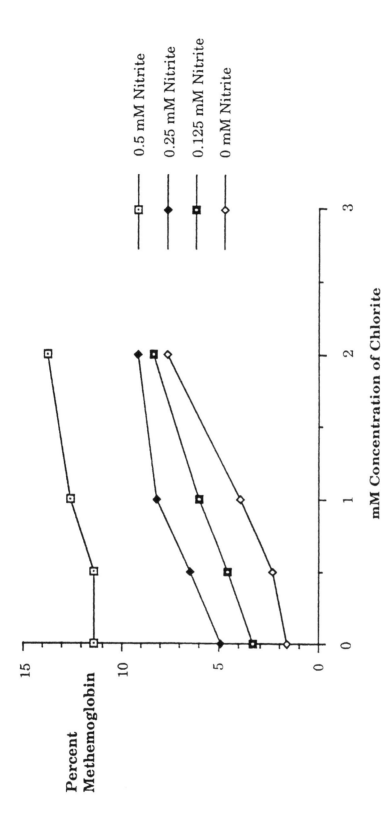

FIGURE **3** : COMBINATION CHLORITE AND NITRITE CONCENTRATIONS
VERSUS PERCENT METHEMOGLOBIN

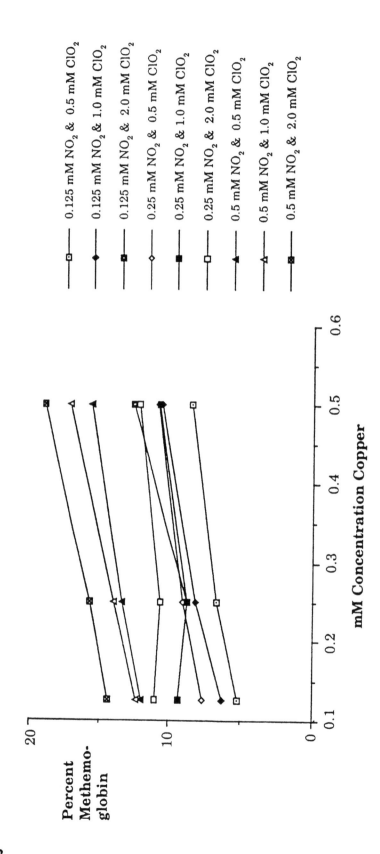

FIGURE 4 : COMBINATION CHLORITE, NITRITE AND COPPER CON-
CENTRATIONS VERSUS PERCENT METHEMOGLOBIN

additivity can be seen in Figure 4. Overall, this response was additive.

DISCUSSION

Because many chemicals must be taken into account when studying environmental contamination, a variety of methods have been proposed for assessing the toxicity of complex mixtures. A common feature of all approaches is that an additive response is usually assumed. For the most part, the data support this assumption. However, in order to determine the type of interaction, the biochemical mechanism of action must be taken into account. The current study was not designed to determine such information.

It is not understood why copper consistantly elicited lower than expected values.

An explanation can be proposed for the one non-additive response of chlorite/nitrite. It could be related to the stability of these ions. Cupric ion and chlorite ion are both relatively stable. Nitrite, on the other hand is not stable. It is readily oxidized to nitrate ion. It is reasonable to assume that chlorite has the ability to oxidize nitrite. Chlorite has what is known as a non-metallic character. The more non-metallic a species is, the greater its tendency to gain electrons and become negative [7]. Metals, on the other hand, have the tendency to remain positive. The chlorine atom is more non-metallic than the nitrogen atom. In light of this chlorine is likely to remove an electron from nitrite even though they have the same oxidation potential. This would reduce methemoglobin formation since nitrate is less potent than nitrite. Chlorite is probably reduced to chloride ion or hypochlorite. The latter agent is capable of producing methemoglobin. This could explain the observed antagonism. The reaction between chlorite and nitrite suggests that these two chemicals would not be found in significant amounts in the same water supply. However, since nitrite and copper have other sources (food, air) it is possible that simultaneous exposure to all three of these chemicals does indeed occur. When all three chemicals were considered, this antagonism seems to have been masked.

In conclusion, this in vitro work shows that the response of combining these methemoglobin-producing agents is additive. The combination of chlorite and nitrite alone produces a less-than-additive response. Generally, though, these preliminary results support EPA's current assumptions about mixtures [8].

REFERENCES

1. Kiese, M. <u>Methemoglobinemia: A Comprehensive Treatise</u>. Cleveland, CRC Press Inc., 1974.

2. Moore, G. and Calabrese, E. The Effects of Copper and Chlorite on Normal and G-6-PD Deficient Erythrocytes. <u>Journal of Environmental Pathology and Toxicology</u>. Volume 4 pp. 271-280, 1980.

3. McGuire and Meadows, 1987, as cited in <u>Health Effects of Drinking Water Treatment Technologies</u>. Drinking Water Task Force. Lewis Publishers, 1989.

4. Lukens, J. The Legacy of Well Water Methemoglobinemia. <u>Journal of the American Medical Association</u>. Volume 257 No. 20, 1987.

5. Smith, J. and Beutler, E. Methemoglobin Formation and Reduction in Man and Various Animal Species. <u>American Journal of Physiology</u>. Volume 210(2) pp. 347-350, 1966.

6. Brown, B. <u>Hematology: Principles and Procedures</u>. Fourth Edition. Philadelphia, Lea and Febiger, 1984.

7. Zaicek, T. Professor of Chemistry, University of Massachusetts, Personal Communication, 1990.

8. USEPA. Guidelines for the Health Risk Assessment of Chemical Mixtures. FRL 2984-2. <u>Federal Register</u>. Volume 51 No. 185, 1986

J. WALTON
P. LABINE
A. REIDIES

The Chemistry of Permanganate in Degradative Oxidations

ABSTRACT

While the chemistry of organic <u>synthetic</u> permanganate oxidations has been studied quite thoroughly over the last 125 years, the chemistry of <u>destructive</u> or <u>degradative</u> permanganate oxidations, especially under the conditions of industrial waste treatment, is still in the process of being developed.

In this paper, the reaction conditions will be characterized under which permanganate will break down specific pollutants or contaminants such as alcohols, olefins, organosulfur compounds and amines. There will also be a discussion of some of the oxidation or breakdown products.

INTRODUCTION

Potassium permanganate ($KMnO_4$) [7722-64-7] has found application in many different fields [1]. In the early eighties potassium permanganate treatment started to become popular as an odor control method in municipal sewage plants and collection lines [2]. Simultaneously it gained in importance as an oxidant for the destruction and/or detoxification of a wide range of environmentally harmful chemicals in industrial waste materials.

Potassium permanganate can also affect TOC, COD, and BOD values. In permanganate oxidations of industrial waste effluents, total organic carbon (TOC) can be lowered by three mechanisms (1) oxidation of organics to CO_2 and H_2O, (2) adsorption of the pollutants and/or their partial oxidation products onto hydrous MnO_2, and (3) enhancing biodegradability.

Chemical Oxygen Demand (COD) will be affected in a similar way. The formation of higher oxygenated organics will be reflected by lower COD values. The effect of pollutant sorption on COD would be the same as for TOC.

J. Walton, P. Labine and A. Reidies, Carus Chemical Company, 1001 Boyce Memorial Drive, Ottawa, Illinois, U.S.A.

Biological Oxygen Demand (BOD) can be affected by permanganate treatment in different ways. In most cases, BOD values will follow the course of COD and TOC (in the direction of lower values) but in some instances, where the oxidation by $KMnO_4$ has broken the organics into more readily biodegradable fragments, BOD values in the treated waste can be somewhat higher than in the untreated waste. Improved biodegradability is generally looked upon as a very desirable treatment result.

PROPERTIES of POTASSIUM PERMANGANATE

PHYSICAL PROPERTIES

Potassium permanganate ($KMnO_4$) occurs either as granular or needle-shaped crystals, both of dark purple color. The density of potassium permanganate is 2.73 g/cm^3, its bulk density is in the order of 100 $lb/ft^3 \approx 1,600$ kg/m^3.

The solubility of $KMnO_4$ in water is strongly influenced by temperature, as shown in Table I.

The heat of dissolution is 10.62 Kcal/mole (44.44 KJ/mole) for dissolving $KMnO_4$ in an unlimited quantity of water and 9.20 Kcal/mole (38.49 KJ/mole) in a (near) saturated solution [3].

Permanganate solutions are intensely colored. Water containing up to a few mg/L will look pink; at higher concentrations, the color is a characteristic purple. The spectrum of permanganate ion contains several maxima between 420 and 650 nm (responsible for the purple color), but the peak most commonly used in spectrophotometric determinations is at 526.5 nm.

TABLE I

Solubility of Potassium Permanganate in Water	
Temperature, degree C	% by Weight
0	2.75
10	4.07
20	5.96
30	8.28
40	11.13
50	24.45
65	20.02

[3] and [4]

CHEMICAL PROPERTIES

Chemistry of Permanganate Oxidations in Aqueous Solutions

Oxidations with permanganate can occur via several different reaction paths: electron abstraction, hydrogen atom abstraction, hydride-ion abstraction, and direct donation of oxygen to the organic substrate [5]. Enthalpies of activation in permanganate oxidations are generally low (frequently in the order of 5-10 Kcal/mole) but entropies of activation are invariably negative. The kinetics of permanganate oxidations are greatly affected by variations in the entropy term, the value of which can be drastically altered by changes in pH. It is largely for that reason that, in the majority of cases, hydrogen ion concentration exerts a powerful influence on reaction rates.

Influence of pH

The vast majority of permanganate oxidations are carried out in aqueous solutions. Under these conditions it is to a large extent the pH of the system that will determine whether the oxidation will involve a one, three, or five electron exchange and whether the reaction will be fast or slow. Under extremely alkaline conditions (pH > 12 to 13), the reduction of the permanganate by the substrate will frequently stop at the manganate (VI) stage, i.e. allow the transfer of only one electron.

$$MnO_4^- + e^- \longrightarrow MnO_4^{2-} \tag{1}$$

In the pH range from approximately 3.5 to 12, permanganate ion will normally undergo a three electron exchange, equation 2a being applicable to acidic pH conditions and equation 2b to alkaline conditions [6].

Potential $E°$
in Volts

$$MnO_4^- + 4 H^+ + 3 e^- \longrightarrow MnO_2 + 2 H_2O \qquad + 1.70 \tag{2a}$$

$$MnO_4^- + 2 H_2O + 3 e^- \longrightarrow MnO_2 + 4 OH^- \qquad + 0.59 \tag{2b}$$

Common to both equations is the formation of manganese dioxide as the reduction product of MnO_4^-, but the initial pH largely determines whether the reaction mixture becomes more alkaline or more acidic in the course of the oxidation [6].

Although the redox potential of permanganate ion decreases with increasing pH, many organic permanganate oxidations proceed faster under alkaline than under neutral or weakly acidic conditions. High pHs favor anion formation, which usually enhances oxidizability over that of neutral molecules. Cations are more oxidation resistant than neutral molecules. Thus, Amines at low pHs are resistant to oxidation due to the formation of ammonium cations [6].

It is useful to remember that even around the neutral pH point (pH 7) permanganate will attack quite a few organic compounds, particularly aliphatic aldehydes and amines, as well as many compounds with olefinic double bonds.

Under more strongly acidic conditions (pH < 3.5) and in the presence of a strong reducing agent, permanganate ion can undergo a five electron exchange, as shown in equation 3:

$$MnO_4^- + 8\,H^+ + 5\,e^- \longrightarrow Mn^{2+} + 4\,H_2O \qquad +1.51 \qquad (3)$$

The practical significance of permanganate reactions under strongly acidic conditions is major for oxidative degradations of difficult-to-destroy-pollutants. In such cases, acidic permanganate can be very desirable because of its high degree of aggressiveness. In particularly stubborn cases, sulfuric acid concentrations can be raised to (30-50% H_2SO_4) where free permanganic acid becomes the predominant oxidizing species. Permanganic acid is a much more powerful oxidant than permanganate ion, but it is also correspondingly more corrosive.

Other Reaction Parameters

Besides pH, permanganate oxidations (rate and degree) are also influenced by temperature, time, and concentrations. Under the conditions of industrial waste treatment, there are often limitations to all three of these parameters. It is usually not economical to perform oxidations that require boiling temperatures in conjunction with many hours of reaction time or that need large stoichiometric excesses of $KMnO_4$.

From a practical point of view, permanganate oxidations should be brought to completion within a few hours at ambient temperatures. Especially in heterogeneous systems, adequate agitation can be of key importance.

Not many permanganate oxidations are known to be affected by catalysts, but heavy metal ions like Cu^{2+} or Ag^+ enhance, for example, the oxidation of cyanide to cyanate. It has also been established that the MnO_2 produced in most permanganate oxidations plays an autocatalytic role in many of these reactions.

OXIDIZABILITY OF COMPOUNDS/FUNCTIONAL GROUPS BY KMnO₄

Some of the functional groups and molecular configurations that are particularly susceptible to permanganate attack will be discussed in this section.

THE CARBON-CARBON DOUBLE BOND

Molecules that contain carbon-carbon double bonds are usually quite readily oxidized by permanganate under rather mild conditions of pH 4-8 and ambient temperature. Under these conditions, α–hydroxyketone is formed [7]. However, if the pH is raised above 9, the diol is the major product. A certain amount of carbon-carbon bond cleavage can also occur but the formation of two carboxylic acids normally occurs under acidic conditions or under conditions of excess $KMnO_4$ and/or higher temperature.

$$
\begin{array}{ccc}
\underset{\text{olefin}}{\overset{\displaystyle \overset{\textstyle H\ \ H}{\overset{\textstyle |\ \ |}{R\text{-}C\text{=}C\text{-}R'}}}{}} & \xrightarrow[\text{pH 4-8}]{\text{KMnO4}} & \underset{\alpha\text{–hydroxyketone}}{\overset{\displaystyle \overset{\textstyle H}{\overset{\textstyle |}{\underset{\overset{\textstyle ||\ |}{O\ OH}}{R\text{-}C\text{-}C\text{-}R'}}}}{}}
\end{array}
\qquad (4a)
$$

$$\begin{array}{ccc}
\underset{\displaystyle\overset{|}{R}\text{-}\underset{}{C}=\underset{}{C}\text{-}R'}{\overset{\displaystyle H \ \ H}{\overset{| \ \ |}{}}} & \xrightarrow[\text{pH} >9]{\text{KMnO4}} & \underset{\displaystyle\underset{\overset{| \ \ |}{HO \ OH}}{\overset{| \ \ |}{R\text{-}C\text{-}C\text{-}R'}}}{\overset{\displaystyle H \ \ H}{\overset{| \ \ |}{}}}
\end{array} \qquad (4b)$$

olefin   diol

Carcinogenic polynuclear aromatic hydrocarbons (PAH) appear to be destroyed by <u>acidic</u> permanganate. Using $KMnO_4$ in aqueous saturated solution or as a 0.3 mol/L solution in 3 molar H_2SO_4, the following PAH were successfully oxidized [8]:

- benzo [α] anthracene [56-55-3]

- benzo [α] pyrene [50-32-8]

- 7,12-dimethylbenzanthrene [57-97-6]

- dibenz[α, α] anthracene [50-32-8]

- 3-methylcholanthrene [56-49-5]

Vinyl chloride [75-01-4] in gaseous or aqueous process streams can be effectively oxidized by $KMnO_4$ at a molar ratio of 1:3.15 [9].

COD reduction by 65-91% was attained by treating wastewater from the production of polystyrene with oxidizing agents such as $KMnO_4$ [10].

ALCOHOLS, ALDEHYDES, AND KETONES

A primary alcohol which contains two α–hydrogens can form aldehydes and carboxylic acids, whereas a secondary alcohol which contains only one α-hydrogen can form ketones. In the presence of excess permanganate, the aldehydes will be converted into carboxylic acids.

$$\underset{\displaystyle\underset{H}{\overset{H}{R\text{-}C\text{-}OH}}}{\overset{| \ }{\overset{| \ }{}}} \xrightarrow{\text{KMnO4}} \underset{\displaystyle R\text{-}C=O}{\overset{\displaystyle H}{\overset{|}{}}} \xrightarrow{\text{KMnO4}} \underset{\displaystyle R\text{-}C=O}{\overset{\displaystyle OH}{\overset{|}{}}} \qquad (5)$$

primary alcohol   aldehyde   carboxylic acid

$$\underset{\displaystyle\underset{H}{\overset{R'}{R\text{-}C\text{-}OH}}}{\overset{|}{\overset{|}{}}} \xrightarrow{\text{KMnO4}} \underset{\displaystyle R\text{-}C=O}{\overset{\displaystyle R'}{\overset{|}{}}} \qquad (6)$$

secondary alcohol   ketone

A ketone will generally be oxidation resistant unless there is enough free alkali present to affect enolization [6].

$$
\underset{\substack{\text{ketone}}}{\underset{\substack{\text{OH}}}{\text{R-C-C-R'}}} \xrightarrow{\text{OH-}} \left[\underset{\substack{\text{enol}\\\text{intermediate}}}{\underset{\substack{\text{O}}}{\text{R-C=C-R'}}} \right]^{-} \xrightarrow{\text{KMnO4}} \underset{\substack{\text{carboxylic}\\\text{acid}}}{\underset{\substack{\text{OH}}}{\text{R-C=O}}} + \underset{\substack{\text{ketone}}}{\underset{\substack{}}{\text{O=C-R'}}}
$$

(7)

The resulting enol will then be oxidatively cleaved into a shorter chain carboxylic acid and a corresponding ketone [6].

Tertiary alcohols will not react with $KMnO_4$ since no α–hydrogens are available.

$$
\underset{\substack{\text{tertiary alcohol}}}{\overset{\text{R'}}{\underset{\text{R''}}{\text{R-C-OH}}}} \xrightarrow{\text{KMnO4}} \text{No reaction} \tag{8}
$$

PHENOLS (Monohydroxybenzene [108-95-2] and Derivatives)

An electron donating hydroxyl group attached to a benzene ring is generally readily degraded by permanganate. In the presence of sufficient oxidant, phenol (hydroxybenzene) is completely converted into CO_2 and H_2O. This occurs at a pH of 7-10 (pH 8.5 to 9.5 preferred) with one to three hours required for complete oxidation. However, almost 90% of the phenol is destroyed during the first 10 minutes [11].

The complete degradation of phenol by permanganate follows this equation:

$$
3\,C_6H_5OH + 28\,KMnO_4 + 5\,H_2O \longrightarrow 18\,CO_2 + 28\,KOH + 28\,MnO_2 \tag{9}
$$

which means that one mg/L of phenol requires 15.7 mg/L of $KMnO_4$.

In practice, less than stoichiometric quantities of permanganate are used, frequently only 6-7 mg/L of $KMnO_4$ per mg/L of phenol. This is adequate in most cases to destroy the phenol as such, but it leaves some degradation products of the phenol intact. The degradation products found in one case include mesotartaric, formic, and oxalic acids [12].

Phenols that have electron withdrawing substituents like halogen or nitro groups on the benzene ring are harder to oxidize than either phenol or any of the cresols (methyl-substituted phenols) [6].

Monochloro- and mononitrophenols respectively will still be degraded by $KMnO_4$ at slower but acceptable rates, but the presence of additional chloro and/or nitro groups on the aromatic ring will render such phenols progressively inert towards oxidation by $KMnO_4$. Thus, pentachlorophenol or picric acid will not be attacked by permanganate under the conditions of industrial waste treatment.

Applications

Phenols occur in the effluents from paper mills, coke mills, oil refineries, thermoplastic resin manufacture, paint stripping operations, foundries, and other manufacturing facilities. Reviews of the phenol problem connected with industrial waste and available solutions are available [13,14]. At concentrations of > 200 ppb phenol, a ratio of 6-7 mg/L of $KMnO_4$ per mg/L of phenol might be adequate. This ratio can become 80:1 or higher at phenol concentrations of < 100 ppb [11]. This is probably because of the presence of other oxidizable constituents at higher concentrations in the waste stream that will compete for the permanganate along with the phenol.

A patented process for the removal of phenol by $KMnO_4$ is described in which the pH is adjusted to >6 and the temperature is held at 30°C. $KMnO_4$ is added at the 70-90% equivalent of the COD to remove phenols and COD [15]. The actual dosage will, of course, depend on the presence of other oxidizable constituents.

Mitin discusses a treatment system in which the waste is treated with 11.5 mg/L of $KMnO_4$ per mg/L of phenol with pH adjustment by addition of CaO. The resulting sludge is first allowed to settle and is then separated by filtration[16].

To remove COD from a wastewater containing phenols and other organics, $KMnO_4$ is added at a dose rate equivalent to 70-90% of the initial COD value [15].

AMINES [6]

When primary aliphatic amines containing α–hydrogens are oxidized, the products of reaction are similar to those with aliphatic alcohols, [17] except that ammonia is also produced. Sec-carbinamines form products similar to those formed by secondary alcohols.

Some typical reactions are shown below.

$$\begin{array}{ccccc} \overset{H}{\underset{H}{R-\overset{|}{\underset{|}{C}}-NH_2}} & \xrightarrow{KMnO4} & \left[\overset{H}{R-\overset{|}{\underset{|}{C}}=\overset{H}{N}}\right] & \xrightarrow{KMnO4} & \overset{H}{R-\overset{|}{C}=O} + NH_3 & \xrightarrow{KMnO4} & \overset{OH}{R-\overset{|}{C}=O} \end{array}$$
$$\text{(10)}$$

primary amine   intermediate   aldehyde ammonia   carboxylic acid

$$\overset{R'}{\underset{H}{R-\overset{|}{\underset{|}{C}}-NH_2}} \xrightarrow{KMnO4} \overset{R'}{R-\overset{|}{C}=O} + NH_3 \qquad (11)$$

sec-carbinamines   ketone   ammonia

Unlike tertiary alcohols, tert-carbinamines react with $KMnO_4$ to form nitrocompounds [6].

$$\underset{\text{tert-carbinamine}}{R-\underset{\underset{R''}{|}}{\overset{\overset{R'}{|}}{C}}-NH_2} \quad \xrightarrow{KMnO_4} \quad R-\underset{\underset{R''}{|}}{\overset{\overset{R'}{|}}{C}}-NO_2 \qquad (12)$$

Secondary amines containing an α–hydrogen are oxidatively hydrolyzed by aqueous permanganate to the corresponding ketones. For example, dicyclohexylamine is converted into cyclohexanone in yields of 70 to 85% [17].

$$\text{(cyclohexyl)}-NH-\text{(cyclohexyl)} \quad \xrightarrow{KMnO_4} \quad 2\;\text{(cyclohexyl)}{=}O \;+\; NH_3 \qquad (13)$$

Simple tertiary amines are oxidized to ketones by use of permanganate in aqueous tert-butanol [17].

$$\text{(cyclohexyl)}_2N-CH_2CH_3 \quad \xrightarrow[CaSO_4]{KMnO_4} \quad 2\;\text{(cyclohexyl)}{=}O \;+\; O{=}\overset{\overset{H}{|}}{C}\text{-}CH_3 \;+\; NH_3 \qquad (14)$$

Primary aromatic amines containing α–hydrogens are generally quite easily oxidized to aldehydes by neutral or basic permanganate. For example, benzylamine ($C_6H_5CH_2NH_2$) and its derivatives are converted to benzaldehyde and substituted benzaldehydes [17].

$$C_6H_5-\underset{\underset{H}{|}}{\overset{\overset{H}{|}}{C}}-NH_2 \quad \xrightarrow{KMnO_4} \quad C_6H_5-\overset{\overset{H}{|}}{C}{=}O \;+\; NH_3 \qquad (15)$$

Secondary aromatic amines are reported to give hydrazones when treated with permanganate ion in aqueous acetone. For example, 2,2'-dinaphthylamine may be converted to tetranaphthylhydrazone in 65% yield [17].

$$Ar-\underset{\underset{H}{|}}{N}-Ar \quad \xrightarrow{KMnO_4} \quad \begin{array}{c} Ar-N-Ar \\ | \\ Ar-N-Ar \end{array} \qquad Ar = \text{(naphthyl)} \qquad (16)$$

212

Aniline and p-phenylenediamine usually form imine and quinone intermediates. In the presence of alkali, the reaction products are azobenzene, NH_3, oxalic acid and other products. Under acidic conditions, aniline black is formed [18].

Other amines are readily oxidizable by neutral and alkaline $KMnO_4$, but to a multiplicity of oxidation products. Protonation in acidic solution will usually decrease the susceptibility of amines to oxidation, as will the presence of carboxylic groups in the same molecule (especially in a conjugate configuration). Complex formation with metal ions like Cr^{3+} and Cu^{2+} can further stabilize the amino group against permanganate oxidation.

Permanganate treatment is also suitable for wastewaters containing odorous nitrogen compounds such as γ-picoline [19] and indole [20]. Diazole dyes have been reported to be attacked by permanganate [21]. Carcinogenic amino and nitro-compounds can be destroyed by $KMnO_4$ in sulfuric acid media. Thus the following amines can be destroyed [22, 23]:

- 4-aminobiphenyl [92-67-1]

- 4, 4'-methylene-bis (o-chloroaniline) [101-14-4]

- 3, 3' diaminobenzidine [91-95-2]

- 1-naphtylamine [134-32-7]

- 2-naphtylamine [91-59-8]

- 2, 4-diaminotoluene [95-80-7]

- N, N-dimethyl-4-amino-4'hydroxyazobenzene [2495-15-3]

The antineoplastic agent Melphalan [148-82-3] (also known as phenylalanine nitrogen mustard) was completely oxidized by alkaline $KMnO_4$ [24].

Potassium permanganate in H_2SO_4 medium was found to destroy N-nitrosamines [25].

SULFUR COMPOUNDS

Organic sulfur compounds such as thiols are rather easily oxidized by permanganate. Aliphatic mercaptans are first oxidized to disulfides and then to sulfonic acids, whereas aromatic thiols are oxidized directly to sulfonic acids.

$$2\,R\text{-}S\text{-}H \xrightarrow{MnO_4^-} R\text{-}S\text{-}S\text{-}R \xrightarrow{MnO_4^-} R\text{-}\underset{O}{\overset{O}{\underset{\|}{\overset{\|}{S}}}}\text{-}OH \qquad R= alkyl \qquad (17)$$

thiol   disulfide   sulfonic acid

(18)

thiol   sulfonic acid

Thioethers and sulfoxides are both oxidized to sulfones.

$$
\begin{array}{c}
\text{R-S-R'} \\
\text{thioether} \\
\\
\text{or} \\
\\
\underset{\underset{\displaystyle O}{\parallel}}{\text{R-S-R'}} \\
\text{sulfoxide}
\end{array}
\quad \xrightarrow{\text{MnO4-}} \quad
\underset{\underset{\displaystyle O}{\parallel}}{\overset{\overset{\displaystyle O}{\parallel}}{\text{R-S-R'}}}
\qquad\qquad (19)
$$

sulfones

Numerous other organosulfur compounds, including sulfinic acids, cyclic sulfites, etc. are also subject to permanganate oxidations [17].

Paper mill sludges can be treated with $KMnO_4$ to control mercaptan odors in the plant atmosphere, specifically in the areas around belt filter presses [26]. A Japanese process [27] recommends air scrubbing with alkaline permanganate to abate mercaptans, thioethers, and H_2S from food processing and/or chemical plants, and subsequent treatment of the spent scrubbing liquor with enough additional permanganate to maintain a redox potential at 520-560 mV.

Sulfur dyes have been reported to be attacked by permanganate [21] and wastes containing the fungicidal ethylenebis-(thiocarbamate) [28141-26-6] are treatable with $KMnO_4$ within the pH range of 1-10 [28].

MISCELLANEOUS OXIDATIONS OF INORGANIC COMPOUNDS WITH $KMnO_4$

Hydrogen Sulfide

Hydrogen sulfide and alkali metal sulfides interact rapidly with $KMnO_4$. Under acidic conditions (pH < 7.5) the predominant reaction can be formulated as:

$$ 3\,H_2S + 2\,KMnO_4 \longrightarrow 3\,S^\circ + 2\,H_2O + 2\,MnO_2 + 2\,KOH \qquad (20) $$

whereas, under alkaline conditions (pH > 7.5) the oxidation mainly proceeds according to:

$$ 3\,H_2S + 8\,KMnO_4 \longrightarrow 3\,K_2SO_4 + 2\,H_2O + 8\,MnO_2 + 2\,KOH \qquad (21) $$

This means that in H_2S-containing wastewaters with pHs around the neutral point, one mg/L of H_2S will require between 3.3 and 13.2 mg/L of $KMnO_4$ [29] depending on which of the two reactions is predominant.

However, field reports (for example Ficek, 1985) indicate that under conditions where there is exposure to air oxygen, permanganate to sulfide ratios as low as 1:1 can suffice to effect the desired odor control [30]. This apparent anomaly can be explained by the catalytic action of the hydrous MnO_2 formed by reduction of the $KMnO_4$ [31].

It must also be considered that the permanganate/H_2S reaction does not only produce elemental sulfur and sulfate but also other sulfur species or intermediate valence stages such as tetrathionate.

Permanganate will also oxidize most other inorganic sulfur compounds in which the sulfur has a valence state of <+6: sulfur dioxide and its salts (bisulfite, sulfite, and metabisulfite), and thiosulfate to name a few.

Cyanides

The permanganate oxidation of CN^- at pH 12-14 leads to the formation of cyanate (CNO^-) ion [32].

$$2 MnO_4^- + 3 CN^- + H_2O \xrightarrow{Ca(OH)_2} 2 MnO_2 + 3 CNO^- + 2 OH^- \qquad (22)$$

If an alkali metal base is used instead of $Ca(OH)_2$, the reduction of the $KMnO_4$ stops at the manganate (VI) stage:

$$2 MnO_4^- + CN^- + 2 OH^- \longrightarrow 2 MnO_4^{2-} + CNO^- + H_2O \qquad (23)$$

Within the pH range of 9-12, the reaction is nonstoichiometric, leading to such oxidation products as CO_2, CNO^- and (internally regenerated) CN^- [33]. At pHs between 6 and 9, toxic cyanogen [$(CN)_2$] is generated.

For the practical detoxification of cyanide ion, only the high pH reaction (eq. 23) is suitable. Copper (II) and silver (I) ions at 2-4 mg/L will significantly enhance the reaction rate. Most complex cyanides (with the exception of the cyano-complexes of iron and silver) are also degraded by $KMnO_4$.

THE ROLE AND SIGNIFICANCE OF HYDROUS MANGANESE DIOXIDE

An important aspect of permanganate waste treatment technology is the formation of hydrous manganese dioxide. Manganese dioxide is the natural end product of MnO_4^- ion reduction under moderately acidic to moderately alkaline conditions.

Hydrous manganese dioxide can catalyze numerous oxidation and decomposition reactions [3] and can autocatalyze many permanganate oxidations of organic compounds [34]. For example, Manganese dioxide will also catalyze the decomposition of organic and other peroxides.

The sorptive properties of hydrous manganese dioxide account for the popular use of $KMnO_4$ in waste treatment applications. Hydrous manganese dioxide (as the name implies) contains varying amounts of chemically bound water, which is largely responsible for the sorptive properties of this amorphous, insoluble, light brown to black colored product. $KMnO_4$ derived hydrous MnO_2 is usually classified as ∂ (delta) manganese dioxide which typically contains adsorbed potassium ion.

There are indications that in the treatment of wastes with potassium permanganate, adsorption on hydrous MnO_2 might be at least partially responsible for those reductions of COD, BOD, and TOC that are in excess of stoichiometrically expected values.

Due to its amphoteric nature, freshly precipitated hydrous MnO_2 can combine with both cationic and anionic species, but it is also capable of scavenging and retaining non-ionic molecules.

OTHER FEATURES AND APPLICATIONS OF $KMnO_4$ OXIDATIONS

COD/BOD REMOVAL

Wastewater from lactam production was subjected to treatment with $KMnO_4$ followed by separation of the sludge (by settling) and passing of the supernatant through a column with activated carbon. The COD was lowered from 946 ppm to 13 ppm [35].

COD reduction by 65-91% was attained by treating wastewater from the production of polystyrene with an oxidizing agent such as $KMnO_4$ [36].

Treating wastewater from epoxy resin production with $KMnO_4$ resulted in COD removal of 65.8% [37].

To remove COD from a wastewater containing phenols and other organics, $KMnO_4$ is added at a dose rate equivalent to 70-90% of the initial COD value [15].

COLOR REMOVAL

Effluents from wood products industries typically exhibit a brown color, due to the presence of saturated and unsaturated organic compounds. This color must largely be removed before the water can be discharged into receiving streams. Biological treatment usually removes only part of the discoloration, the remainder must be destroyed by chemical means. Permanganate treatment is one of the options for this decolorization step [38].

Color problems are also encountered with effluents from food, chemical, and textile processing plants. A process for the decolorization of caramel wastewater using permanganate or alternative oxidants was described by Hidaka [39].

EDTA OXIDATION

Waste from electroless plating operations contains metals (Ni, Cu) as well as chelating agents, such as EDTA (ethylenediaminetetraacetic acid). Before the metals can be precipitated as either hydroxides or sulfides, the chelants present must be oxidatively destroyed. Acid permanganate (pH 1-2.5) has found widespread use for this purpose. Some of these installations are well automated with regard to $KMnO_4$ addition (redox potential) and pH adjustment [40,41,42]

DYE INDUSTRY APPLICATIONS

Wastewaters containing dyes occur in the textile and other industries. Whether or not or to what degree a particular dye will be destructed by $KMnO_4$ will largely depend on its chemical structure, but such dye classes as active, diazoles, vat, and sulfur dyes have been reported as being attacked by permanganate [21].

There is also a patent by Dolya and others covering the use of permanganate for the treatment of dye-containing wastewaters from an automotive plant [43].

DETOXIFICATION

Potassium permanganate can be used to detoxify a range of complex organics including specific pesticide residues. The following organophosphorus insectides have been found degradable by potassium permanganate:

- Dimethoate [60-51-5] also know as Cygon® (American Cyanamid) or Rogor [44]

- Guthion® (Mobay) [86-50-0] also known as Gusathion [45].

- Chlorfenvinphos [470-90-6] also known as Shell 4072; Bromfenvinphos [33399-00-7]. These two pesticides, when present in water at a concentration of 50 mg/L, are completely decomposed by approximately 0.3 g/L $KMnO_4$ within 12 hours [46].

- Parathion [56-38-2] is oxidized at pH 9 mainly to p-nitrophenol; however, under neutral or acidic conditions the predominant oxidation product is still very toxic paraoxon [311-45-5] [47].

- The bipyridilium herbicide Paraquat [4685-14-7] is detoxified by $KMnO_4$ at neutral or slightly acidic pHs [47].

- Wastes containing the fungicidal ethylenebis-(thiocarbamate) [28141-26-6] are treatable with $KMnO_4$ within the pH range of 1-10 [28].

- Permanganate is widely used to detoxify the fish toxicants rotenone [83-79-4] and antimycin [27220-56-0] [48].

EMULSION BREAKING

Wastes containing oil emulsions (spent cutting oils, degreasing baths) are treated with acidified $KMnO_4$ at 50 to 90° COD to destroy the emulsifier and thus to break the emulsion. Separated oils are skimmed off the top. The process is repeated until the content of extractable substances in the aqueous phases reaches acceptable levels [49,50].

DETERGENT REMOVAL

Effluents from a cosmetics plant were retreated with alum as flocculating agent, $KMnO_4$ and $Ca(ClO)_2$ as oxidizers, and activated carbon as an absorbent. This treatment is said to remove all detergents and 95% of all other pollutants [51].

CONCLUSION

Potassium Permangante is a strong oxidant that has found many uses in industrial waste water treatment. It has the ability to oxidize alcohols, olefins, organosulfur compounds, amines, and other compounds. Potassium Permanganate also helps to lower TOC, COD, and BOD values.

Alcohols can be oxidized with $KMnO_4$. However, these reactions may be slow. Primary alcohols are oxidized to aldehydes and carboxylic acids depending on the ratio of $KMnO_4$ to alcohol. Secondary alcohols are usually oxidized to ketones. Tertiary alcohols will not react. Phenol and substituted phenols can be partially oxidized to mixtures of carboxylic acids or completely oxidized to carbon dioxide if the waste water matrix does not have an extremely high $KMnO_4$ demand.

Olefins are oxidized to α–hydroxyketones or diols depending on pH conditions. Carbon-carbon bond cleavage can also occur under more severe conditions.

Organosulfur compounds such as thiols, thioethers, and sulfoxides are also quite reactive with $KMnO_4$. Alkylthiols are oxidized to disulfides whereas arylthiols are oxidized to sulfonic acids. Thioethers and sulfoxides are oxidized to sulfones.

Inorganic sulfides and cyanides are also oxidized easily. Inorganic sulfides such as hydrogen sulfide are oxidized to elemental sulfur or sulfate depending on the reaction conditions. Inorganic cyanides such as sodium cyanide are oxidized to cyanate.

Manganese dioxide, one of the by-products of potassium permanganate reactions is also useful in applications in terms of its adsorption and catalytic properties.

REFERENCES

1. Reidies, A. Water Cond. Purif., Oct 1987, p. 32-33.

2. Pisarczyk, K and Rossi, L. A., "Sludge Odor Control and Improved Dewatering with Potassium Permanganate", paper presented at the 55th Annual Conference of the Water Pollution Control Federation , St. Louis MO, 5 Oct 1982.

3. Gmelin Handbook of Inorganic Chemistry, 8th Edition. Manganese Part C 2 (in German) System Number 56 Springer-Verlag, Berlin-Heidelberg-New York 1975.

4. Reidies, A. H. Kirk-Othmer Encyclopedia of Chemical Technology, Vol 14, Third edition, p. 844-895, Copyright 1981 by John Wiley and Sons, Inc.

5. Stewart, R. Oxidation Mechanisms. Applications to Organic Chemistry, W. A. Benjamin, Inc., New York-Amsterdam 1964.

6. Stewart, R. Oxidation in Organic Chemistry, Vol 5 A, Academic Press, New York - London, p. 2-85 1965.

7. Fatiadi, A.J. Journal of Synthetic Organic Chemistry, Vol. 2, 86-88, 1987.

8. Castegnaro, M., Coombs, M., Phillipson, M.A., Bourgade, M.C., and Michelon, J. Polynucl. Aromat. Hydrocarbons, Int. Symp., 7th 257-68 (Eng) 1982 (Publ. 1983). CA 99: 200007p.

9. Witenhafer, D. E., Daniels, C. A. and Koebel, R. F. U.S. 4,062,925, 13 Dec 1977. CA 88: 54620n.

10. Masat, J. Czech. CS 211,451, 15 Apr 1984. CA 102: 666925m.

11. Throop, W.M. Journal of Hazardous Material 1(4) 319-329 1977.

12. Bobkov, V. N. Tr. VNII Vodosnabzh., Kanaliz. Gidrotekhn. Sooruzh. i Inzh. Gidrogeol. (50), 48-50 (Russ) 1975.
CA 86: 21516p.

13. Lanouette, K. H. Chem. Eng. 17 Oct 1977.

14. Throop, W. M. and Boyle, W. C. Proc. Annu. Pollut. Control Conf. Water Wastewater Equip. Manuf. Assoc., 3rd 115-43 (Eng) 1975.
CA 84: 155259z.

15. Atsuchi K., Hagiwara, E. and Fujita, K. Japan. Kokai 78 30, 163, 22 Mar 1978.
CA 89: 152073b.

16. Mitin, B.A., Popyrina, L.M., and Rabinovich, A.L. Okhr. Prir. Vod Urala 8, 53-7 (Russ) 1975.
CA 87: 43623u.

17. Lee, D. The Oxidation of Organic Compounds by Permanganate Ion and Hexavalent Chromium, Open Court Publishing Company 1980.

18. Beilstein, Vol. 12, p. 67. 1929.

19. Saenko, S. I., Nosikova, T.A., Zakhorzhevskaya, A. G.and Kovaleva, L. A. U.S.S.R. SU 1,038,295, 30 Aug 1983.
CA 99:181054h.

20. Taran, P. N. and Kasyanchuk, R. S. Nauchn. Osn. Tekhnol. Obrab. Vody, 2, 74-7 (Russ) 1975.
CA 88:67755u.

21. Venediktova, K. P. and Utkina, A. M. Gig. Tr. Prof. Zabol. (2), 8-12 (Russ) 1978.
CA 89: 11310q.

22. Barek, J. Microchem. J. 33(1), 97-101 (Eng) 1986.
CA 104: 173841u.

23. Barek, J. and Kelnar, L. Microchem. J. 33(2), 239-42 (Eng) 1986.
CA 104: 229892f.

24. Barek, J., Castegnaro, M., Malaveille, C., Brouet, I. and Zima, J. Microchem. J. 36(2), 192-7 (Eng) 1987.
CA 108: 43336e.

25. Castegnaro, M. Michelon J. and Walker, E.A. IARC Sci. Publ. 41(N-Nitroso Compd: occurrence Biol. Eff.), 151-7 (Eng) 1982.
CA 99:10297k.

26. Jackson, J.H. Pulp Pap. 58(4), 147-9 (Eng) 1984.
CA 101: 27695w.

27. Ohbiki, T., Ueno, K., Negoro, S., Wariishi, H., Sugimoto, N. Japan. Kokai 77,147,574, 8 Dec 1977.
CA 89: 64446v.

28. Kuchikata, M. and Kuyama, H. Japan. Kokai 78 21,080, 27 Feb 1978.
    CA 89: 117279b.

29. Cadena, F. and Peters, R.W. J. Water Pollut. Control Fed. 60(7), 1259-63
    (Eng) 1988.
    CA 109: 79071z.

30. Ficek, K. Unpublished report: "Permanganate Controls Sewage Odors", Carus
    Chemical Company, LaSalle, IL (Form #310) 1985.

31. Eye, D.J. and Clement, D.P. J. Amer. Leather Chem. Assoc. 67(6), 256-67 1972.
    CA 77: 79239f.

32. Posselt, H. S. Journal of Organic Chemistry. 37, 2763-65, 1972.

33. Stewart, R. and Van der Linden, R. Can. J. Chem. 38, 2237-55, 1960.
    CA 55: 10179d.

34. Garrido, J. A., Perez-Benito, J. R., Rodriguez, R. M., de Andres, J. and Brillas, E. J.
    Chem. Res. (S), 380-81 1987.

35. Munechika, K., Shimosaka, M. and Tajiri, H. Japan. Kokai 73 69,359, 20 Sep 1973.
    CA 80: 40788w.

36. Masat, J. Czech. CS 211,451, 15 April 1984.
    CA 102: 66925m.

37. Sharifov, R. R. and Bondareva, N.I.Azerb. Kim. Zh. (3), 119-22 (Russ) 1981.
    CA 96: 148460c.

38. Davis, E. and Hwang, C.P. Waste Treat. Util., Proc. Int. Symp., 217-29 (Eng)
    1978 (Pub. 1979).
    CA 95: 85722w 1979.

39. Hidaka, M., Ide, T., Izumida, K. and Yamaguchi, T. Shokuhin Kogyo
    16(22), 89-96 (Japan) 1973.
    CA 80: 63595v.

40. Schroeter, J. and Painter, C. "Potassium Permanganate Oxidation of Electroless Plating
    Wastewater" , National Conference of Water Pollution Control Federation, Los
    Angeles, CA. 5-6 Oct 1986.

41. Feikes, L., Schroeder, D. and Reyer, P. Ger. Offen. DE 3,335,746, 11 April 1985
    CA 103: 26768b.

42. Inakuma, Y. Japan Kokai Tokyo Koho 78,133,962, 22 Nov 1978.
    CA 90: 60846y.

43. Dolya, I.M., Sitnik, A.S., Yankovskii, E.R., and Chechel, N.V.   U.S.S.R. SU
    882,949, 23 Nov 1981.
    CA 96: 129191z.

44. Kravets, E.V. Nauchn. Osn. Tekhnol. Ochisttki Vody, 46-7 (Russ) 1973. CA 82:76933v.

45. Chambon, P. Geoffray, C. and Vial, J. Bull. Trav. Soc. Pharm. Lyon, 17(2), 57-8 (France) 1973. CA 80: 63552d.

46. Drabent, Z., Gosiewskia, H., Krasnicki, K., and Malinowska, B. Zesz. Probl. Postepow Nauk Roln. 319, 347-55 (Pol) 1986. CA 106: 182049k.

47. Gomaa, H. M. and Faust, S.D. Advan. Chem. Ser. 111 (Fate of Org. Pestic. in the Aquatic Environ.), 189-209 (Eng) 1972. CA 77: 160881f.

48. Duncan, Th. O. " A Review of the Literature on the Use of Potassium Permanganate in Fisheries" report No. FWS-LR-74-14, PB-235-453 distributed by NTIS. 1974.

49. Reichman, F. and Heller, Z. Czech. CS 237,038, 15 Feb 1987. CA 107: 120531y.

50. Reichman, F., Ernest, J. and Heller, Z. Czech. CS 236,995, 1 Oct 1986. CA 107: 120526a.

51. Catalan Lafuente, J.G., Bustos Aragon, A., Mora Duran, J., and Cabo Ramon, J. Doc. Invest. Hidrol. 18, 271-84 (Span) 1974. CA 85: 112339d.

PHILIP A. VELLA*
JOSEPH A. MUNDER

Uses of KMnO$_4$, H$_2$O$_2$ and ClO$_2$ in Wastewater Applications

ABSTRACT

Wastewater containing phenolic and reduced sulfur species pollutants is an area of growing concern. Each year, industrial discharge permits are becoming more stringent relative to the release of phenolic compounds into the environment. To reach these new phenol limits (μg/L), closer examination of the available oxidation technologies is needed. This study was undertaken to evaluate the effectiveness of potassium permanganate, hydrogen peroxide and chlorine dioxide for the oxidation of low level phenols. The goal of this initial study was to determine optimum dose, pH and the effects of calcium and phosphate on phenol oxidation. The data was obtained from in-house laboratory testing in a "pure" water system and two industrial wastewater samples. Evaluation of these oxidants on a sulfidic containing wastewater will also be described.

Results of the "pure" water investigation indicated that ClO$_2$ was the most versatile oxidant under the conditions studied. It was not affected by pH, calcium or phosphate ion and required the lowest dose on a w/w basis for phenol oxidation. Potassium permanganate was also very flexible but did show a pH dependency near 9. Due to pH, hydroxyl radical scavenger effects, and iron chelates, Fenton's system was the most limited oxidant studied. Examination of these oxidants in wastewater matrixes showed significant differences from the predicted results based on the "pure" water data. These differences can be attributed to diversity of the waters studied.

BACKGROUND

CHEMICAL OXIDATION OF PHENOLS

Literature and practice support a number of methods for the treatment of phenolic wastewaters. A common treatment is chemical oxidation. Listed in Table I is a comparison of some commonly used oxidants.

* TO WHOM ALL CORRESPONDENCE SHOULD BE ADDRESSED

Philip A. Vella and Joseph A. Munder, Carus Chemical Company, 1001 Boyce Memorial Drive, Ottawa, Illinois, U.S.A.

TABLE I. CHEMICAL OXIDANT OPTIONS

Oxidant	Complete Oxidation	Oxidant:Phenol Ratio w/w	pH	Catalyst Required	Treatment Form
O_3[1,2]	yes	2-6:1	7-12	no	Dissolved gas*
Cl_2[3]	yes	100:1	7-8	no	Liquid or gas**
ClO_2[1,4]	yes	1.5-3.5:1	2-10	no	Dissolved gas*
H_2O_2[1,5]	yes	2-5:1	4-6	yes	Liquid
$KMnO_4$[1,6]	yes	16:1	7-10	no	Crystal

*Generated On-Site
**Hypochlorite may be applied

Lanouette[1] reports that ozone will oxidize phenol (1.5-2.5 w/w ratio) over a broad pH range but is phenol selective at 11.5 to 11.8. At low phenol levels, the quantity of ozone required to oxidize phenol is greater than 2.5:1. Ozone requires on site generation and has a high capital cost.

Throop and Boyle[7] demonstrated a 100:1 chlorine to phenol ratio for complete oxidation, and a recommended pH range of 7-8.3 to minimize chlorophenol formation. Chlorine and hypochlorite are inexpensive but may present control problems in the avoidance of chlorinated phenols formation.[1,7]

Lanouette[1] reports that chlorine dioxide at 1.5 times the amount of phenol (pH 7-8) will oxidize phenol to benzoquinone and at a 3.3 ratio (pH 10), phenol is reportedly oxidized to maleic and oxalic acids.[1] Furthermore, he reports that chlorine dioxide will not produce chlorophenols and will oxidize any chlorophenols present.[1] Chlorine dioxide is also generated on site. It is produced as a gas and is explosive at levels above 10% in air. However, upon generation it is immediately dissolved in water to form a safe and stable solution.[2,4] It works very effectively on low phenol concentrations and becomes cost prohibitive at high phenol levels.

Throop also provided data on the iron salt catalyzed oxidation of phenol by hydrogen peroxide. A ratio of approximately 2:1 hydrogen peroxide to phenol (w/w) is required to achieve greater than 99% phenol oxidation.[1] Hydrogen peroxide is effective at high phenol concentrations and marginal at low levels. The effective pH range of peroxide is limited to between 3-5.

Potassium permanganate has been widely used for the oxidation of phenolic compounds. In 1973 Kroop[8] reported the results of a limited set of laboratory experiments on the oxidation of phenols in paint stripping wastewater. Throop and Boyle[7] and Rosfjord[9] and coworkers published comparisons of phenol removal methods including the use of potassium permanganate. Complete oxidation requires a KMnO4 to phenol ratio of 15.7:1 but reports indicate that a 6-7:1 ratio will afford ring cleavage[10]. KMnO4, being a readily soluble solid, presents materials handling advantages resulting in minimum installation costs. In addition it does produce MnO_2 as an oxidation by product which can be useful in the adsorption of organics.

ALTERNATIVE OXIDANTS

Other phenol treatment options including physical and biological processes may be applied to the reduction of phenol concentrations in wastewater streams. As early as 1975[11] studies demonstrated that activated carbon removes phenols down to several parts per billion in the effluent. Biological removal of phenols has also been recommended.[12] The effluent limitations for direct dischargers, using end-of-pipe biological treatment, are more restrictive (26 µg/L daily maximum and 15 µg/L maximum monthly average)[12] than those applied to non-biological treatment systems.

EXPERIMENTAL

REAGENTS

Reagents were of technical grade or better. The hydrogen peroxide was 35% technical grade. The water used in all tests was softened, deionized followed by reverse osmosis resulting in water with a conductivity of approximately 15 megohm/cm . Stock solutions were prepared in class A volumetric flasks. Stock solutions of phenol were stored in amber bottles at 4° C.

PREPARATION OF A CHLORINE DIOXIDE SOLUTION

Chlorine dioxide was prepared by the reaction of sodium chlorite with sodium hypochlorite under acidic conditions. To a brown bottle containing 500 ml of water, 30 ml of a 5% sodium hypochlorite solution were added (ClO_2 decomposes under sunlight). To this solution 4.2 ml of concentrated hydrochloric acid were added. After mixing, 5 ml of a 25% sodium chlorite solution were added and mixed thoroughly. The concentration of this ClO_2 stock solution was determined spectrophotometrically at 445nm in a 1 inch cell using a HACH DR-2000 spectrophotometer.

GENERAL TEST CONDITIONS

All test solutions were stirred at the rate of 100 rpm. The pH was adjusted to 4, 7 or 9 with ammonium hydroxide or sulfuric acid. Except where noted, reaction times were 15 minutes. Testing was done at 1 and 10 mg/L phenol concentrations. At concentrations greater than 1 mg/L, the reactions were quenched with sodium thiosulfate. Below this level, no quenching was done due to the method of analysis used (see below). All ratios are reported as w/w unless noted.

PHENOL ANALYSIS

High Level Analysis (>1 mg/L)

Phenol concentrations greater than 1 mg/L were analyzed spectrophotometrically using a modified 4-aminoantipyrine procedure[13]. This procedure involves filtering the sample through a 0.45µm filter, mixing 50 mL of the filtrate with 1 mL of pH 10 buffer, followed by 1 ml of a 20 g/L 4-aminoantipyrine solution, then 1 mL of a 80 g/L potassium ferricyanide solution. Samples were swirled between additions. The resulting solution was analyzed at 510 nm in a 1 inch cell using a HACH DR-2000 spectrophotometer. A standard curve was generated from 0.5 to 3 mg/L phenol. Periodic standards were run to verify the curve.

Low Level Analysis (< 1 mg/L)

At phenol concentrations less than 1 mg/L, the samples were analyzed using High Pressure Liquid Chromatography (HPLC). The analysis followed the procedure described in the DIONEX methods handbook[14] using an Ion Pac NS-1 column. Samples were filtered through a 0.2μm filter prior to injection on the instrument. When using the HPLC, no sodium thiosulfate was added to quench the reactions because thiosulfate has been shown to depress the sensitivity of the detector. Therefore, analysis was done within 1 minute of the set reaction time to insure comparable results.

TOTAL REDUCER ANALYSIS

A sample of wastewater is titrated with a standard solution of iodine until an iodine residual, as indicated by the blue I_3^- starch complex, is attained. The amount of iodine consumed in the titration indicates the amount of readily oxidizable pollutants present. Substances registering as Total Reducers (TR) include inorganic sulfides, sulfite, thiosulfate, sulfur dioxide, mercaptan, nitrite, free cyanide, reduced forms of soluble metals and certain unsaturated organic compounds. Using 0.01N Iodine solution, TR can be calculated as shown below.

$$TR \text{ (mg/L as } I_2) = 1270 \times B/A \quad \text{where} \quad A = \text{amount of titrant used} \qquad (1)$$
$$B = \text{sample aliquot in mls}$$

RESULTS AND DISCUSSION

LABORATORY STUDIES IN A "PURE" WATER MATRIX

Reaction with Potassium Permanganate.

The ratios of permanganate to phenol examined were 1:1, 4:1, 7:1, and 8:1. Figures 1 and 2 illustrate the results of these tests.

The reaction between 10 mg/L phenol and potassium permanganate was most effective at an 8:1 ratio (Fig 1). At this ratio, oxidation was 97-99% complete. With this same ratio and 1 mg/L phenol, oxidation was only 66% complete (Fig 2). It has been shown in a number of applications when low levels of phenol are to be oxidized, more than the theoretical amount of $KMnO_4$ is required.[15]

At a concentration of 10 mg/L phenol, no pH effect was observed. At 1 mg/L, a pH of 9 proved to be the most effective.

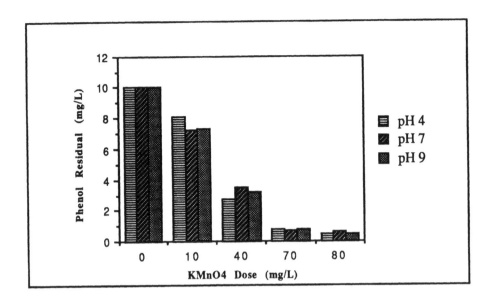

Figure 1 Effects of pH and Concentration; Oxidation of 10 mg/L Phenol with KMnO$_4$

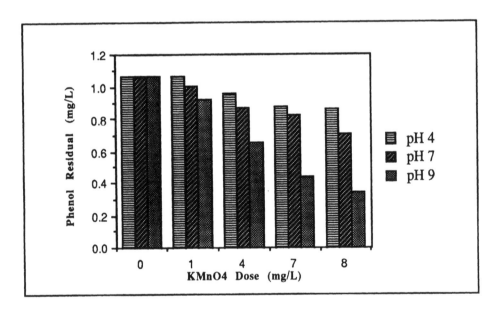

Figure 2 Effects of pH and Concentration; Oxidation of 1 mg/L Phenol with KMnO$_4$

226

Effect of Phosphate

The effect of phosphate on the oxidation of phenol was examined. The $KMnO_4$ to phenol ratio was 4:1. Phosphate, in the form of NaH_2PO_4, was added in concentrations of 50, 100, and 200 mg /L. Adjustment to pH 9 was made after phosphate was added.

Figures 3 and 4 depict the effect of phosphate on the permanganate system. From Figure 3 there was a small improvement in the oxidation of 1 mg/L phenol at pH's of 4 and 7 (8 and 3% respectively) with a phosphate concentration of 200 mg/L/. Figure 4 shows that addition of phosphate slightly hinders the oxidation of 10 mg/L phenol at pH 7 and 9 with a greater effect observed at a pH of 4.

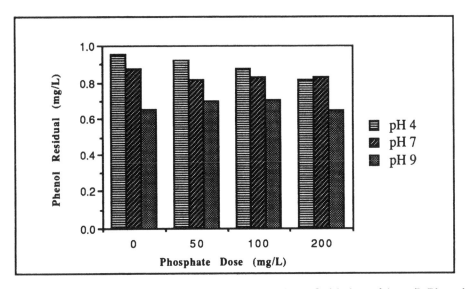

Figure 3 Effect of pH and Phosphate Concentration; Oxidation of 1 mg/L Phenol with 4 mg/L $KMnO_4$

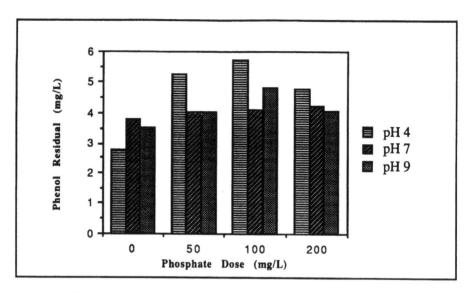

Figure 4 Effect of pH and Phosphate Concentration; Oxidation of 10 mg/L Phenol with 40 mg/L $KMnO_4$

Effect of Calcium

It has been theorized that calcium is an aid to the oxidation process of permanganate. To verify this, the effects of calcium were investigated. Calcium, as $CaCO_3$, was added to DI water which was acidified to drive off the CO_2. Phenol was added and the pH adjusted.

Figures 5 and 6 depict the effect of calcium. From Figure 5 there was a 18% increase in phenol oxidation of 1 mg/L phenol with the addition of 200 mg/L Ca at a pH of 9. Figure 6 illustrates no significant effect at pH of 7 and 9 and a moderate decrease in oxidation at a pH of 4. The increased oxidation may be due to calcium aiding in the electron transfer from the manganate 6+ to the manganese dioxide 4+ state. This would provide more efficient use of the oxidizing power of $KMnO_4$. There is no current agreement for the observed decrease in oxidation as the $KMnO_4$ concentration is increased.

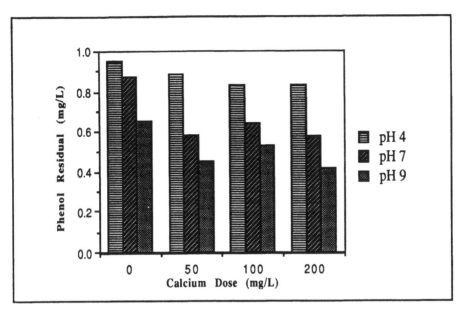

Figure 5 Effect of pH and Calcium Concentration: Oxidation of 1 mg/L Phenol
with 4 mg/L KMnO$_4$

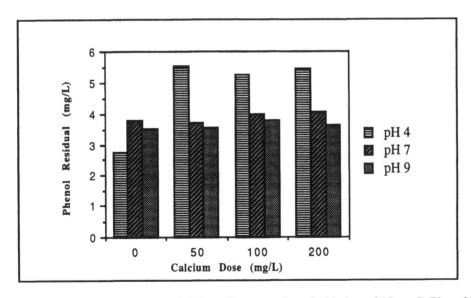

Figure 6 Effect of pH and Calcium Concentration: Oxidation of 10 mg/L Phenol
with 40 mg/L KMnO$_4$

Reaction of Phenol with Catalyzed Peroxide (Fenton's System)

Tests were carried out at 1 & 10 mg/L phenol levels. All pH adjustments were made with dilute H_2SO_4 and NaOH. Iron catalyst was added as $FeSO_4$ in a 1:10 ratio to peroxide. During testing, the iron was added first and allowed to mix well before peroxide was added. Peroxide concentrations were at ratios of 1:1, 4:1, 6:1, and 8:1 to phenol. Due to the pH dependency of this system (OH· generation is efficient at low pH's), only pH 4 was investigated.

The optimum dose was 6 mg/L H_2O_2 for the 1 mg/L phenol solution and 40 mg/L H_2O_2 for the 10 mg/L solution. (See Fig. 8) There is no graph for the oxidation of 10 mg/L phenol because there was only one point (the oxidation was complete in the remaining tests). Using this pH and dose, the effects of phosphate and calcium were evaluated. In addition, an experiment was performed to test if permanganate could be substituted for the iron catalyst. The results of this test were negative.

Figures 7-9 show the results of the catalyzed peroxide testing. In a 1 mg/L phenol solution, oxidation was 98% complete with a peroxide dose of 6 mg/L after 15 minutes. After 30 minutes, the phenol was oxidized to below the detection limit (0.005 mg/L). The 10 mg/L phenol solution required only a 40 mg/L peroxide dose to oxidize the phenol to below detection limits in 15 minutes. Due to rapid oxidation, no graphs were generated at this phenol level.

Addition of NaH_2PO_4 (Fig. 8 and 9) to the Fenton system had a negative effect. At 1 mg/L phenol the oxidation dropped from nearly 100% to near 30%. When added to the 10 mg/L phenol system, the oxidation was reduced to 15%. This result is due to the chelation of the iron by the phosphate ion. With the iron bound, the generation of free radicals is severely limited. Another property of the phosphate ion is as a free radical trap. Although not the primary reason in this case, this will also have a negative effect on free radical attack of organics.

As tested here, there was no effect of calcium ion and no graphs were generated. This indicates that calcium in not an effective radical trap. If the carbonate had been allowed to remain in solution, a negative would be expected since carbonate is a known free radical trap.

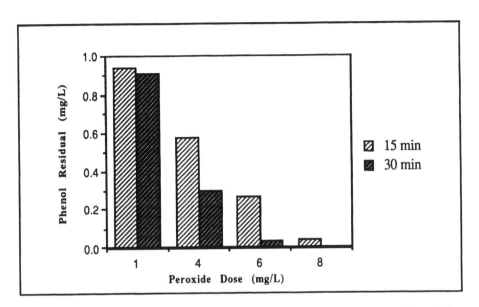

Figure 7 Effects of H_2O_2 on Fenton's System; Oxidation of 1 mg/L Phenol at pH 4

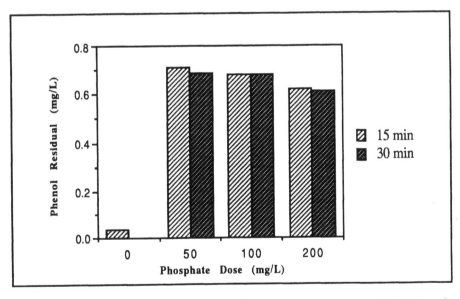

Figure 8 Effects of Phosphate Concentration on Fenton's System; Oxidation of 1 mg/L Phenol with 4 mg/L H_2O_2 at pH 4

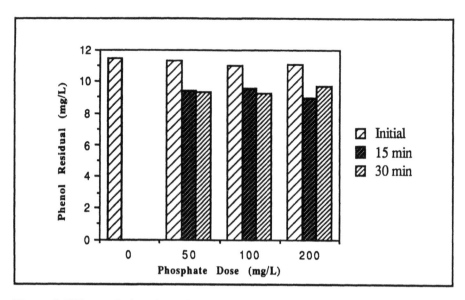

Figure 9 Effects of Phosphate Concentration on Fenton's System; Oxidation of 10 mg/L Phenol with 40 mg/L H_2O_2 at pH 4

Reaction of Phenol with Chlorine Dioxide

Chlorine dioxide was reacted at 0.5, 1, 2, and 4 times the phenol concentration. The pH ranges investigated were 4, 7 and 10. Due to the instability of the stock solution, the chlorine dioxide concentration was determined prior to each test . As with previous oxidants, the effects of phosphate and calcium were examined.

Figures 10 and 11 present the results from the chlorine dioxide experiments. Chlorine dioxide proved to be a very effective oxidant, even in the presence of the additives. In a 1 mg/L phenol solution, pH 10 proved to be the best for oxidation. At this pH, the minimum concentration of chlorine dioxide resulting in 100% oxidation was 2 mg/L. Similar results were also observed at pH 7.

In a 10 mg/L phenol solution there was very little pH effect. A 4:1 dose of chlorine dioxide to phenol exhibited the best oxidation (91%) in a 10 mg/L phenol solution.

Addition of NaH_2PO_4 at all concentrations had no adverse effect on the oxidation of the phenol at either 1 mg/L or 10 mg/L levels. Results from the addition of Ca were identical.

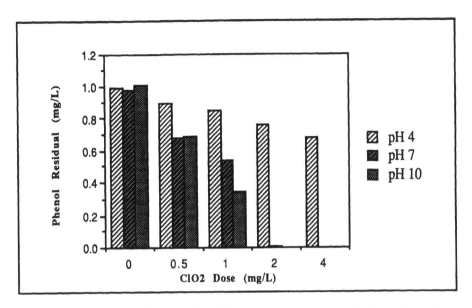

Figure 10 Effects of Chlorine Dioxide Concentration and pH; Oxidation of
1 mg/L Phenol

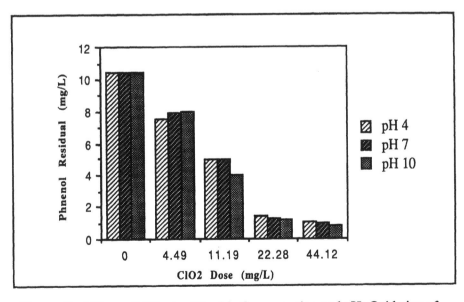

Figure 11 Effects of Chlorine Dioxide Concentration and pH; Oxidation of
10 mg/L Phenol

Wastewater phenol samples from a metal caster ere obtained for an oxidant treatability study. A sample from a steel foundry was obtained for sulfidic oxidation. The results of the treatability studies are given below.

Metal Caster

The data obtained in this study is presented in Table II. Figures 12 and 13 illustrate the phenol residual after treatment with ClO_2, and $KMnO_4$. Results indicate that chlorine dioxide at 8 mg/L effectively reduced phenol levels to below the target of 0.4mg/L. Potassium permanganate at 30 mg/L was also effective. Fenton's system at dosages up to 200 mg/L H_2O_2 (not shown, phenol residual 0.81 mg/L) proved to be relatively ineffective. This is most likely due to the high concentration of iron in the water which decomposed the peroxide and limited the formation of hydroxyl radicals. Other factors such as OH radical scavengers could also limit oxidation.

TABLE II: OXIDANT DOSE AND PHENOL RESIDUAL

Oxidant	Dosage (mg/L)	Phenol (mg/L)
Untreated		0.97
ClO2	5.3	0.90
	20.4	0.08
KMnO4	10.0	0.76
	39.8	0.27
	59.6	0.11
	99.0	0.11
Fenton's	200.0	0.81

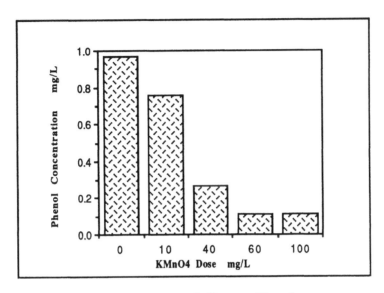

Figure 12 Metal Caster; KMnO$_4$ Dose vs Phenol

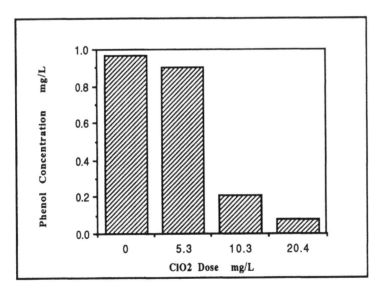

Figure 13 Metal Caster; ClO$_2$ Dose vs Phenol

The results of the oxidant study comparing $KMnO_4$, H_2O_2 and ClO_2 are presented in Figures 14-16. As seen, amounts in excess of 8,000, 17,000 and 5,100 mg/L of $KMnO_4$, H_2O_2 and ClO_2 respectively are required to bring total reducer levels from approximately 25,000 mg/L to near 0 mg/L. A cost comparison of these options indicates that $KMnO_4$ and H_2O_2 are nearly equivalent with ClO_2 40% higher.

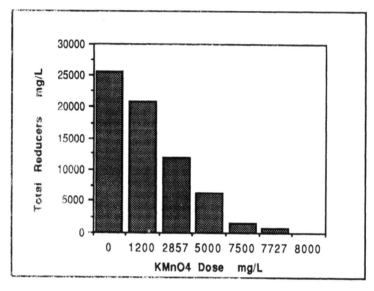

Figure 14 Steel Foundry; $KMnO_4$ Dose vs Total Reducers

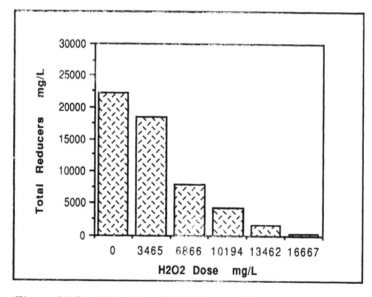

Figure 15 Steel Foundry; H_2O_2 Dose vs Total Reducers

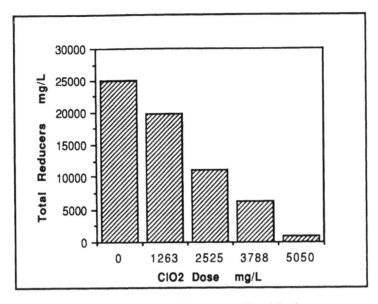

Figure 16 Steel Foundry; ClO$_2$ Dose vs Total Reducers

CONCLUSIONS

PURE WATER STUDIES

1) Chlorine dioxide was the most flexible oxidant studied. It was not affected by the introduction of phosphate or calcium ion and performed well at all pH ranges studied. It also required the lowest dose on a w/w basis for phenol oxidation.

2) The capability of potassium permanganate closely followed that of ClO$_2$. At low phenol concentrations, a pH dependency was observed with pH of 9 being optimum. It was enhanced by calcium ion at low concentrations. At higher concentrations, calcium and phosphate had a negative effect at a pH of 4.

3) Iron catalyzed peroxide (Fenton's) performed well but is pH dependent (at pH of 4), and is susceptible to iron chelates and hydroxyl radical scavengers (PO$_4$, CO$_3$).

WASTEWATER STUDIES

Since wastewaters are extremely diverse, no prediction on oxidant efficiencies can be made without individual testing. In the case of the metal caster, both ClO$_2$ and KMnO$_4$ were effective for phenol oxidation while Fenton's system was ineffective.

For the oxidation of reduced sulfur species in the steel foundry, the oxidation capacity of ClO$_2$ and H$_2$O$_2$ are substantially different on a weight bases. However, when determining treatment cost, each oxidant became equal.

REFERENCES

1.   Lanouette, K.H., "Treatment of Phenolic Wastes", Chemical Engineering Deskbook Issue, pp. 99-106, October 17, 1977.

2.   Conway, R.A. and Ross, R.D., Handbook of Industrial Waste Disposal, pp. 224-225, Van Nostrand Reinhold Company, New York, 1980.

3.   Throop, W.M., "Alternative Methods of Phenol Wastewater Control", J. of Hazr. Mater., pp. 319-329, 1(4), 1977.

4.   Masschelein, W. J., Chlorine Dioxide Chemistry and Environmental Impact of Oxychlorine Compounds, Ann Arbor Science, Ann Arbor MI 1979

5.   Kibbel, W.H., Jr., "Peroxide Treatment for Industrial Waste Problems", Industrial Water Engineering, pp. 6-11, August/September, 1976.

6.   Bobkov, V. N., Tr. VNII Vodosnabzh., Kanaliz, Gidrotekhn. Sooruzh. 1 Inzh. Gidrogeol. pp. 48-50, (50), 1975, (Russ.).

7.   Throop, W. M. and Boyle, W. C., "Perplexing Phenols Alternative Methods for Removal", Proc. Annu. Pollut. Control Conf Wastewater Equip. Manuf. Assoc., 3rd 1975, pp. 115-143.

8.   Kroop, R. H., "Treatment of Phenolic Aircraft Paint Stripping Wastewater", Engineering Bulletin of Purdue University, Engineering Extension Series, No. 142, pt. 2, pp. 1071-1087, 1973.

9.   Rosfjord, R. E., "A Water Pollution Control Assessment", Water and Sewage Works, pp. 96-99, March, 1976.

10. Humphrey, S. B., and Eikleberry, M. A., Unpublished Internal Report, Carus Chemical Company, Ottawa, Il., 1962.

11. Patterson, J. W., Wastewater Treatment Technology, p. 207, Ann Arbor Science, Ann Arbor, MI, 1975.

12. Newton, J., OCPSF "Effluent Limits, Pretreatment, and NSPS", pp, 48-51. Pollution Engineering, March 1988.

13. Standard Methods for the Examination of Water and Wastewater, 16th edition, pp. 556-560, American Public Health Association American Water Works Association, Water Pollution Control Federation, Washington, D. C., 1985.

14. Dionex Application Update, Dionex Corporation 1988.

15. Carus Internal Reports.

C. P. HUANG
CHIEH-SHENG CHU

Electrochemical Oxidation of Phenolic Compounds from Dilute Aqueous Solutions

ABSTRACT

The feasibility of electrochemical oxidation of selected organic compounds was evaluated in the laboratory using cyclic voltammetric techniques. A rotating ring disk (Pt) electrode was used. The phenolic compounds of environmental importance were selected for investigation. The results show clearly that phenolic compounds are electrochemically oxidizable, as reflected by the unique characteristics of voltammograms. The anodic current is proportional to the removal rate of organic substances. It also noticed that the aliphatic compounds are found in the products of electrolysis of phenols. There is strong evidence that aromatic ring cleavage is achieved at the electrode surface. Among the phenolic compounds tested, the phenols with the most amount of hydroxy functional group are the easiest to be oxidized; the greater the number of hydroxy group, the larger the anodic current. The other factor, which is also important in affecting the oxidation efficiency, is the pH value in the solution. Under the alkaline condition, the hydroxyl ions are oxidized to form the hydroxyl radicals in the anodic region. Those radicals would react with the neutral molecules and cause the benzene ring to undergo hydrolization.

Keywords: Electrochemical oxidation; Phenols

I. Introduction

In the United States, groundwater has always been a major water resource for many domestic, industrial and agricultural activities --- about half of the population rely on subsurface water as sole drinking water source. The protection of this most valuable natural resource is of the utmost importance. However, groundwater systems are subjects for contamination by a wide variety of hazardous chemicals, namely heavy metals and toxic organic compounds. Most of these hazardous chemicals have originated from past poorly designed and managed landfills and waste chemical disposal operations.

The U. S. Environmental Protection Agency has publicized over 32,000 localities nationwide as toxic waste sites (the Superfund sites). As many as 2,000 of these sites pose severe health hazards to the

C. P. Huang and Chieh-sheng Chu, Department of Civil Engineering, University of Delaware, Newark, Delaware, U.S.A.

general publics[1]. In the mid-Atlantic area, there are 153 toxic waste sites. The motivation of this research is to study a new physical-chemical process, electrochemical oxidation, for the in-situ removal of toxic organic chemicals from contaminated groundwater. The process studied is primarily intended for emergency groundwater contamination clean-up operations. Nevertheless, the process can also be used wholly or partially for daily groundwater purification and treatment of leachate or high strength industrial wastewater.

Electrochemical method, e.g. cathodic protection, has been used to control corrosion problem in underground pipes and structures for some time. Electro-osmosis has been suggested as a means for the dewatering and consolidation of foundation for construction purposes[2-5], although some the concentrations of some compounds would slightly increase after the electrolysis process[6]. Moreover, electro-deposition technique has been used in the treatment of plating wastes. With careful selection of system parameters and design, this same principle can be applied to in-situ treatment of contaminated groundwater.

II. Materials and Methods

Chemicals

Phenolic compounds used were purchased from either J. T. Baker Chemical Company or Aldrich Chemical Company with a purity of 98% or greater. All chemicals were used as received.

Voltammetry Experiments

The rotating ring disk electrode (model AFDTI36PTPTT platinum) was obtained from Pine Instrument Company, Grove City, Pennsylvania. Apiezon grease (Fisher scientific, Pittsburgh, Pennsylvania) was used to seal the disk into place. This allows the oxidation reaction only occur on the disk surface and to prevent any seeping of the electrolyte into sides of the disk. The resistance between the ring and disk was checked to insure that the two circuit would not become connection inside the electrode. The reference electrode was a saturated calomel electrode (SCE) (Fisher Scientific Company, Model 13-620-52). The counter electrode was a platinum wire electrode (Pine Instrument Company, Grove City, Pennsylvania, Model RR137665A). **Figure 1** shows the experimental setup of the three-electrode voltammetric system. Those three electrode were connected to a Model EVA-19-2 Polarography/Potentiostat which controlled the applied voltage and measured the responsive current. Current and potential were recorded on a Moseley (Model 7001A) x-y recorder purchased from Hewlet Packard Company.

The reaction vessel was double-jacketed for temperature control. The volume of the beaker is about 600 mL. The temperature was controlled by a temperature controller from B. Braun Company (Model Frigomix 1495). The temperature was controlled at 25°C. The electrolyte was 0.1 M KCl (Fisher Scientific Company) or 0.1 M NaClO$_4$ (Alfa Chemical Company) for aqueous solutions and 0.1 M TBAP (tetra butyl ammonia perchloride) for nonpolar organic solvent and the solvent for nonpolar organic solute was acetonitrile (SIGMA Company).

A pH controller was used to control the pH during electrolysis. The controller was purchased from New Brunswick Scientific Company, Edison, New Jersey. The applicable magnitude of for a RDE is in the range $100 < \omega < 1000$ rpm. Each organic compound has different voltammogram. According to its unique voltammetric characteristics, the appropriate oxidation potential was determined. The degradation rate is related to the anodic current. Direct analysis of phenol concentration was done with uv spectrophotometry (Hitachi-Perkin Elmer UV-VIS spectrophotometer) at a wavelength of maximum absorption.

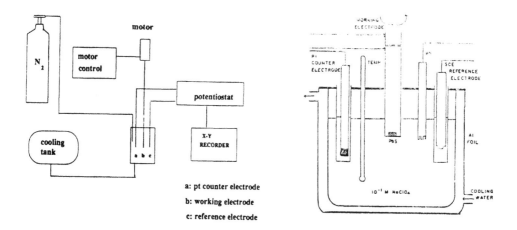

Figure 1. Experimental Setup of 3-Electrode Voltammetry.
Disk outer radius = 0.500 cm. Ring inner radius = 0.750 cm.
Ring outer radius = 0.850 cm.
Thickness of gap = 0.125 cm.
Disk area 0.196 cm^2. Ring area = 0.126 cm^2.
Whole length = 18.57 cm.

III. Results and Discussion

Voltammograms were recorded to study the electrochemical behavior. It is a plot of potential (volt vs. SCE) and current. There are three electrodes in the cell (working, counter, and reference electrode). When the potential of the working electrode reaches a certain value, a current is observed which is resulted from the redox reaction of substances on the electrode surface.

Most of the phenolic compounds with carboxylic group or another hydroxyl group can be readily oxidized on the Pt-electrode. Phenolate anion and phenoxium cation were formed during the electrolysis process. Those ions would react with water or solvent medium which leads to a protonation or deprotonation reaction[7]. Phenolate anion and phenoxium cation might also react with each other to form the enolate compound[8]. Chemical reactions with solvent were preceded by the electron transfer and formed the intermediates which can trigger second electron transfer leading to ring cleavage[9]. The compound with benzene ring structure would transfer to an aliphatic chain structure. The final products have lower free energy, less toxicity and easier biodegradation[10].

a. Oxidation of Phenolic Compounds

Most of the organic compounds found in the groundwater belong to the aromatic group[11]. The phenolic compounds studied are: (a) phenols and like compounds, (b) aromatic compounds, and (c) simple carboxylic acid. Phenols, and their derivatives are the group of organic that receive most attention in oxidation studies.

Results indicate that phenolic compounds are oxidizable, as reflected by their unique voltammetric characteristics (**Figure 2**). It must be mentioned that the pH of the solution also plays a very important role in anodic oxidation. Usually, in alkaline solution, the anodic currents are much higher and last longer than under acid condition. The color of the solution become darker after electrolysis in the alkaline range, due probably to the dimerization of the phenolic compounds. The dimerization reaction is irreversible, inspite of the attemp to adjust the pH back to the original values. The voltammograms are distinctly different before and after electrolysis. Therefore, it is important to realize the optimal pH for the oxidization reaction. **Figure 3** summaries the effect of pH on the oxidation of phenolic compounds.

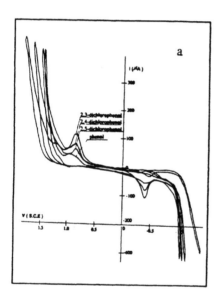 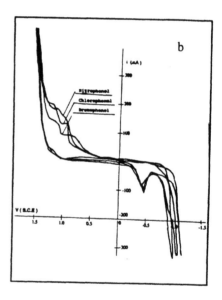

Figure 2.   Voltammograms of Phenols. (a) Chlorophenols, (b) Substituted phenols.
Experimental conditions: concentration of organic 10^{-3} M;
ω = 1,000 rpm; KCl = 10^{-1} M; sweeping rate = 33.3 mv/sec;
temperature = 25 C; pH = 11.

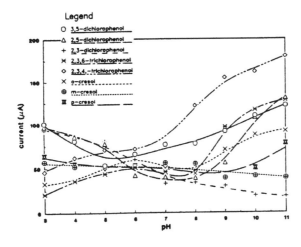

Figure 3. Effect of pH on the Anodic Oxidation of Phenols.
Experimental conditions: concentration of organic 10^{-3} M;
ω = 1,000 rpm; KCl = 10^{-1} M; sweeping rate = 33.3 mv/sec;
temperature = 25 C.

b. Oxidation of Organic Acids

The redox behavior of salicylic acid (2-hydroxy benzoic acid), protocatechuic acid (2,3 - dihydroxybenzoic acid), and gallic acid (3,4,5- trihydroxy-benzoic acid) were also investigated. **Figure 4** summarizes the anodic current as a function of pH for these acids. It is concluded that the more hydroxy groups on the benzene ring, the larger the anodic current. This means that hydroxyl groups are more easily decomposed electrochemically. The current of the anodic oxidation is proportional to the number of hydroxyl group on the benzene ring. Examination the voltammogram of salicylic acid, protocatechuic acid, and gallic acid (at pH 11), the following order can be established: trihydroxy- (gallic) > dihydroxy- (protocatechuic)- > monohydroxy (salicylic)-benzoic acid (**Figure 5**).

This above observation can be attributed partly to the increasing electron withdrawing power of the hydroxy functional group. The hydroxy group and chloride are electronphilic. They would extract the electron density from the benzene ring, stabilize the phenoxenium[12], and finally increase the anodic current[13-15].

243

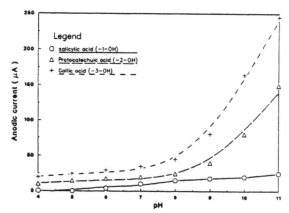

Figure 4. Effect of Hydroxyl Group on Anodic Oxidation.
Experimental conditions: concentration of organic 10^{-3} M;
ω = 1,000 rpm; KCl = 10^{-1} M; sweeping rate = 33.3 mv/sec;
temperature = 25 C.

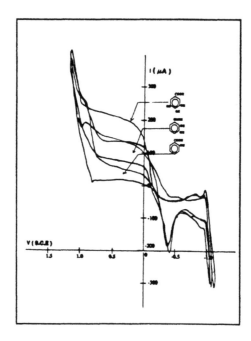

Figure 5.   Voltammogram of Trihydroxyl-, Dihydroxyl-, Monohydroxyl-Benzoic
Acid.
Experimental conditions: concentration of organic 10^{-3} M;
ω = 1,000 rpm; KCl = 10^{-1} M; sweeping rate = 33.3 mv/sec;
temperature = 25 C; pH = 11.

A comparison of non-carboxyl phenolic compounds with the hydroxyl group yields same results: their anodic currents following the order: pyrogallol (3OH) > catechol (2OH) > phenol (OH) (**Figure 6**). This sequence can be described by the theory of electrode polarography developed by Nicholson and Shain[16]. According to these authors, the chemical reaction could couple with the electron transfer[12]. At pH > pK_a, the phenolic molecules become phenolate,

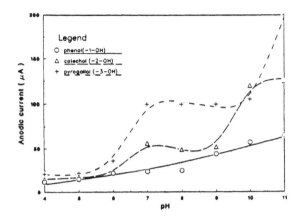

Figure 6.   Relationship between the Number of Hydroxyl Group on Benzene Ring and Anodic Current.
Experimental conditions: concentration of organic 10^{-3} M; $\omega = 1,000$ rpm; KCl = 10^{-1} M; sweeping rate = 33.3 mv/sec; temperature = 25 C.

which is oxidized to release CO_2 and phonxy radical. This is the well known Kolbe reaction. The phenoxy radical can be oxidized again to form phenoxenium cations. The stability of this cation depends on the electron density of the benzene ring. The electronphilic functional group such as hydroxy, chloride, bromide, would draw the electrons out of the benzene ring and decrease the electron density. This means that the cations are stabilized by these substitutions leading to a larger anodic current. Similar results are also shown in the dichlorophenol group: o-dichlorophenol > m-dichlorophenol > p-dichlorophenol (**Figure 7**). In the alkaline condition, the anodic current is larger than the lower pH region.

The phenoxium cation can be continually reduced to 1,4-cyclodihexene-5-ketone. This compound can be protonated with water and becomes 1,4 cyclodiexanol, an intermediate. The final product is tartaric acid. This reaction takes place most favorable at pH larger than 9. After the electrochemical process, the voltammogram is almost identical to tartaric acid at alkaline conditions. Tartaric acid is an aliphatic compound of the TCA cycle. Therefore, it is a strong evidence that the ring cleavage is possible in the electrochemical system.

Based on results obtained, it is possible to propose a reaction mechanism for the electrochemical oxidation of selected organic compounds.

Proposed Electrochemical Oxidation Mechanism.

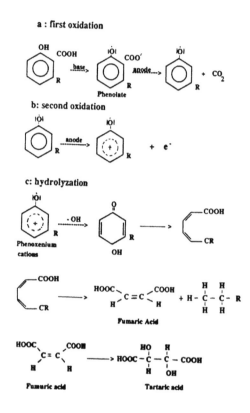

At the first oxidation peak, the aromatic compounds undergo an electron transfer reaction to form phenolate which then becomes phenolic anion while yielding CO_2. This is a typical reaction of the Kolbe type. As anodic oxidation proceeds to the more anodic level, i.e. second peak, the phenolic anion becomes phenoxenium cation, loosing one electron. Under alkaline pH condition, hydrolyzation reaction takes place to convert the phenoxenium cation to quonolic compounds. Further oxidation reaction produces intermediates such as fumaric acid and tartaric acid. The production of tartaric acid was verified by GC-MS analysis of the product. As shown in **Figure 8**, the GC-MS signals clearly indicate the production of tartaric acid.

Results from voltammetric studies, also yield information on the reaction mechanism and the thermodynamic information of the selected organic compounds. Table 1 summarizes the measured standard EMF, $E°$, values for the various organic compounds.

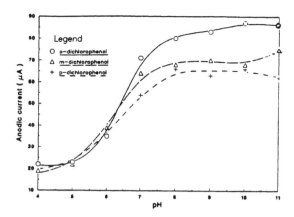

Figure 7.   Relationship between Different Position of Dichlorophenol and Anodic Current.
Experimental conditions: concentration of organic 10^{-3} M;
ω = 1,000 rpm; KCl = 10^{-1} M; sweeping rate = 33.3 mv/sec;
temperature = 25 C.

Table 1.   Summary of Half-wave Potential, Standard EMF, and Reaction Mechanism of Electrochemical Oxidation of Phenols.

Compound	$E_{1/2}$(V vs SCE)	E_p (V vs SCE)	E^0(V vs SCE)	MECHANISM
Phenol	0.633	0.671	0.640	E_rC_i
o-Chlorophenol	0.625	0.556	0.621	ECE
m-Chlorophenol	0.734	0.812	0.785	ECE
p-Chlorophenol	0.653	0.614	0.681	ECE
o-Nitrophenol	0.864	0.898	0.873	E_rC_r
m-Nitrophenol	0.855	0.850	0.824	E_rC_i
p-Nitrophenol	0.924	0.911	0.886	E_rC_i
o-Cresol	0.556	0.551	0.557	C_rE_r
m-Cresol	0.607	0.562	0.547	C_rE_i
p-Cresol	0.543	0.547	0.542	C_rE_i
catechol	0.349	0.354	0.347	E_rC_r
resorcinol	0.613	0.576	0.541	E_rC_r
hydroquinone	0.534	0.577	0.542	C_rE_r
pyrogallol	0.622	0.741	0.612	E_rC_i
benzonic acid	0.451	0.374	0.428	E_rC_i
salicylic acid	0.845	0.766	0.748	ECE
protocatechuic acid	0.994	1.022	0.999	ECE
gallic acid	0.674	0.711	0.681	ECE
aniline	0.700	0.625	0.692	E_rC_r
2-hydro 5-nitro benzonic acid	1.255	1.156	1.136	ECE

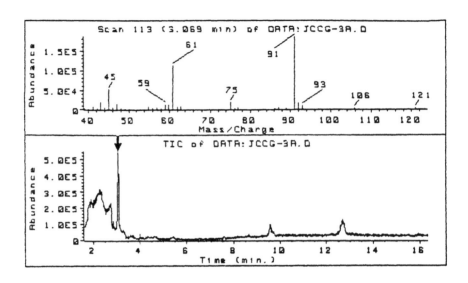

Figure 8.   GC-MS Analysis of By-Products from Constant Potential Electrolysis of Garlic acid.
Experimental conditions: concentration of organic 10^{-3} M; ω = 1,000 rpm; KCl = 10^{-1} M; sweeping rate = 33.3 mv/sec; temperature = 25 C.

c. Oxidation of Surfactant

In a separate study, the oxidation of a surfactant, octanoic acid pentadeca-fluoro-ammonium ($C_7F_{15}COONH_4$), was conducted. **Figures 9** shows the voltammograms of the surfactant over a 16 hours period. After the first cyclic voltammetric experiment, the solution was placed under a constant voltage, 1.12 volt vs. SCE for electrolysis. At the end of the electrolysis, the pH was recorded and the solution was scanned again with the cyclic voltammetric apparatus. Five hours later, the pH changed from original 10.5 to 9.5 and the anodic current peak dropped from 205 to 95 μA. Nine hours after, the pH decreased from 9.5 to 8.9 and the anodic current dropped from 95 to 66 μA. Sixteen hours after, the pH decreased from 8.9 to 8.4 and the anodic current dropped from 66 to 14 μA.

A second experiment was conducted to investigate the oxidation after ester hydrolization. Ester hydrolization can render surfactant decomposed into perfluoro-alkyl-carboxylic acid and ammonium. Since the product, perfluoro alkyl-carboxylic acid, is hydrophobic, the surface tension will increase upon decomposition of the surfactant. As a result of the decomposition reaction, the concentration of ammonium will also increase, and the pH will decrease. The results obtained agree with what is expected. A solution of 200 ppm surfactant was prepared and heated at constant at 95 C for four hours. The solution was then subjected to electrolysis. Immediately after the first

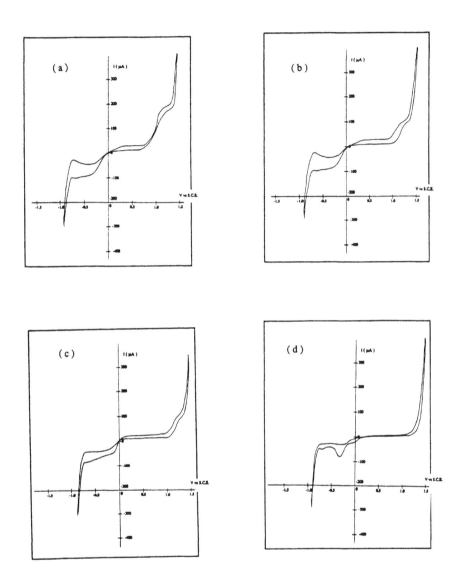

Figure 9. Voltammogram of Surfactant.(a). Initial; (b).4 hr; (c).8 hr; (d). 29 hr.
Experimental conditions: concentration of surfactant = 200 ppm;
ω = 1,000 rpm; KCl = 10^{-1} M; sweeping rate = 33.3 mv/sec;
temperature = 25 C.

electrolysis cycle, the pH dropped from 10.5 to 10.2. The anodic potential shifted from 1.12 to 1.06 volt vs. SCE. But the peak anodic current increased from 205 to 650 µA. This indicated a significant increase in the oxidation rate as reflected by the increase in the anodic current. The solution was then electrolyzed at 1.06 volt vs. SCE for an extended period of time and the pH and ammonium concentration in the solution measured. **Figure 10** shows the changes in pH (Figure 10a), in the ammonium concentration (Figure 10b), in surface tension (Figure 10c) and anodic current during 16 hours constant potential electrolysis.

From the results presented, it is clear that, electrochemically, the surfactant can be broken to ammonium, carbon dioxide and a hydrophobic dimmer, $C_{14}F_{30}$, which is no longer soluble in water. Note that the surface tension of water is 70.9 dyne/cm. The present of surfactant has decreased the surface tension of water from 70.9 to 33.2 dyne/cm. As surfactant is decomposed, the surface tension of the surfactant solution approached that of the water. From the results shown above, the following reaction schemes are proposed:

$$CF_3-(CF_2)_6-\overset{O}{\overset{\|}{C}}-O-NH_4 \quad \xrightarrow{\;H^+\;} \quad CF_3-(CF_2)_6-\overset{\overset{\displaystyle H}{|}\;\overset{\displaystyle O}{|}}{C}-O-NH_4$$

$$CF_3-(CF_2)_6-\overset{\overset{\displaystyle H}{|}\;\overset{\displaystyle O}{|}}{C}-O-NH_4 \quad \xrightarrow{\;+H_2O\;} \quad CF_3-(CF_2)_6-\overset{\overset{\displaystyle H}{|}\;\overset{\displaystyle O}{|}}{\underset{\underset{\displaystyle H\quad H}{\overset{+}{O}}}{C}}-O-NH_4$$

$$CF_3-(CF_2)_6-\overset{\overset{\displaystyle O}{\|}}{\underset{\underset{\displaystyle H\quad H}{\overset{+}{O}}}{C}}-O-NH_4 \quad \xrightarrow{\;rearrangement\;} \quad CF_3-(CF_2)_6-\overset{\overset{\displaystyle OH}{|}\;\overset{\displaystyle H}{|}}{\underset{\underset{\displaystyle OH\quad NH_4}{|\quad\;|}}{C-O}}{}^{+}$$

$$CF_3-(CF_2)_6-\overset{\overset{\displaystyle OH}{|}\;\overset{\displaystyle H}{|}}{\underset{\underset{\displaystyle OH\quad NH_4}{|\quad\;|}}{C-O}}{}^{+} \quad \xrightarrow{\;-ROH\;} \quad CF_3-(CF_2)_6-C\overset{\overset{+}{O}\,H}{\underset{O\,H}{}} \quad + \quad NH_4OH$$

$$CF_3-(CF_2)_6-C\overset{\overset{+}{O}\,H}{\underset{O\,H}{}} \quad \xrightarrow{\;Base\;} \quad CF_3-(CF_2)_6-C\overset{O}{\underset{O\,H}{}} \quad + \quad NH_4OH$$

$$CF_3(CF_2)_6-C\overset{O}{\underset{OH}{}} \quad + \quad NH_4OH \quad \xrightarrow{\;Base\;} \quad C_{14}F_{30} \quad + \quad CO_2 \quad + \quad NH_4OH$$

250

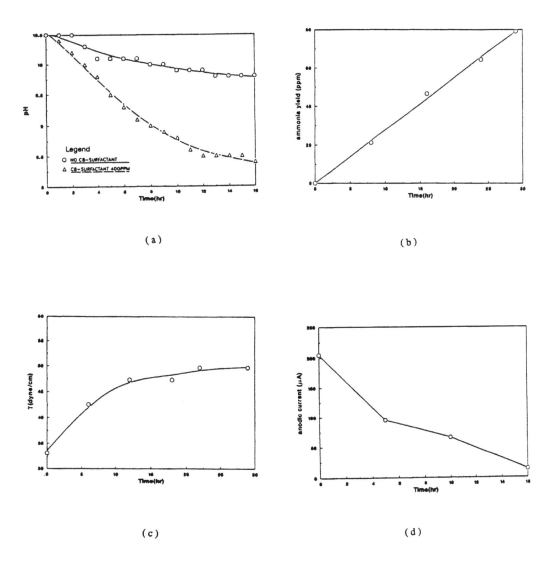

Figure 10. Changes of (a) pH, (b) Ammonium and (c) Surface Tension, and (d) Anodic Current of Surfactant Solution Under Electrolysis. Experimental conditions: sweeping rate = 33.3 mv/sec; rotating speed = 1,000 rpm.

IV. Summary

Results obtained from this study clearly indicate that phenolic compounds are electrochemically oxidizable. This is demonstrated by the unique voltammetric characteristics. The anodic current is proportional to the removal rate of organic compounds. Electrolysis of phenols yields aliphatic compounds. There is strong evidence to demonstrate that aromatic ring cleavage take place electrochemically. Among the phenolic compounds tested, the phenols with the most number of hydroxy functional group are the easiest to be oxidized; the greater the number of hydroxy group, the larger the anodic current. This is due to the increase in the electronphilic substitution which reduces the electron density of benzene ring and stabilizes the intermediate cation in the reaction. Therefore, it enhances the anodic current.

The other factor, which is also important to the oxidation efficiency is the pH value in the solution. This is especially important among the electrochemical intermediates that are coupled with protonation or deprotonation reaction. Under the alkaline condition, the hydroxyl ions are readily oxidized to hydroxyl radicals in the anodic potential region. Those radicals would react with the neutral molecules and cause the benzene ring to undergo hydrolization. Therefore, the efficiency is always higher under alkaline conditions than under the acid conditions. Reversely proton would react with carbonyl group (aldehyde) and proceed with a acid-catalyzed reaction. The product, hemiacetal, would stop the electrochemical reaction.

The final product of electrolysis has the voltammogram same as tartaric acid. It is a aliphatic compound that does not have benzene ring any more. This is a strong evidence that the phenols ring cleavage can be achieved by electrochemical process. Tartaric acid also can be found as an intermediate in the biochemical pathway of the metabolism of aromatic compounds. Therefore, electrochemical process coupled with biological degradation is promising in complete decomposition of phenolic compounds in the groundwater.

V. Acknowledgment

The research on which this report is based was financed in part by the United States Department of the Interior as authorized by the Water Resource Research and Development Act of 1978 (P.L. 95-467). Contents of this publication do not necessary reflect the views and policies of the United States Department of the Interior, nor mention of trade names or commercial products constitute their endorsement by the U. S. Government.

V. References

1. Pojasek, J. W., Membrane Ultrafiltration Disposing of Hazardous Chemical Waste, Environ. Sci. & Tech.,13(7), 810, 1979.

2. U. S. Environmental Protection Agency, National Priorities List, Final, and Proposal Sites, Superfund Office, Washington, DC, July 1989.

3. Bjerrum, L., Moum, J. and Eide, O.," Application of Electro-Osmosis on a Foundation Problem in Norwegian Quick Clay", Geotechnique, 17, No.3, 214-235, 1967.

4. Chappell, B. A., Burton, P. L. "Electro-Osmosis Applied to Unstable Embankment", <u>Journal of Geotechnical Engineering, ASCE</u>, Vol 1, 101, No. GT8, pp 733-740.

5. Casagrande, L., "Electro-Osmosis in Soil", <u>Geotechniqe</u>, vol. 1, pp.1959-1977, 1949.

6. Staas, E. B., Waste Disposal Practices. A Threat to Health and The Nation' Water Supply, Report to the U. S. Congress, CED-28-120, GAO, June 16, 1978.

7. Hoigne, J. <u>Handbook of Ozone Technology</u>, vol.1, Ann Arbor Sci., MI, 1981.

8. Bauch, H and Burchard H., <u>Wasser Luft und Betrieb</u>, <u>14</u>, 270, 1970.

9. Kalovda, R., "Oxidation of Organic compounds with Electrolytically Generated Oxidants", <u>J. Electroanal Chem Interfacial Electrochem</u>, <u>24</u>, 53, 1970.

10. Mallevialle, J., "Measurements of Surface Water Pollution", <u>T. S. M. l'Eau, 70</u>, 107, 1975.

11. Eisenhauer, H. R., "The Ozonation of Phenolic Wastes", <u>J. Water Poll'n Cont'l Fed.</u>, <u>40</u>, pp.1116-1128, 1964.

12. Speiser, B., Rieker, A.,"Electrochemical Oxidations. Part IV. Electrochemical Investigation into the Behavior of 2,6-di-tert-butyl-4-(4-dimetyyl aminophenyl)-phenol. Part I. Phenol and the Species Derived from It: Phenoxy Radical, Phenolate Anion, and Phenoxenium Cation", <u>J. Electroanal. Chem</u> 1102, 373-395, 1979.

13. Baizer, M. N. <u>Organic Electrochemistry</u>, Marcell Dekker, NY, 1985.

14. Inoue, H. and Shikata, M, "Catalytic Action on Cyclhexanol Methylcyclohexanol and Its Derivatives", <u>Japan Ind Chem Soc.</u>, <u>24</u>, 567, 1921.

15. Miller, L., "Anodically Initiated Aromatic Iodination and Side Chain Substitution Reactions", <u>Tetrahedron Letter</u>, 1831, 1968.

16. Nicholson, R. S., and I. Shain, "Theory of Stationary Electrode Polargraphy", <u>Anal. Chem.</u>, <u>36,</u> 706, 1964.

17. Solomon, T. W. Grams, <u>Fundamentals of Organic Chemistry</u>, John Wiley & Sons, New York, NY 1982.

HUSSAIN AL-EKABI*
GAIL EDWARDS
WENDY HOLDEN
ALI SAFARAZADEH-AMIRI
JOAN STORY

Water Treatment by Heterogeneous Photocatalysis

ABSTRACT

The TiO$_2$ photocatalytic degradation of mixtures of o-chlorophenol (o-CP), 2,4-dichlorophenol (2,4-DCP), 2,6-dichlorophenol (2,6-DCP) and 2,4,6-trichlorophenol (2,4,6-TCP) were examined in the presence and absence of electron acceptors. Electron acceptor additives such as hydrogen peroxide, ammonium persulphate and oxygen were chosen for this study. These additives markedly increased the degradation rates of the chlorinated phenol mixtures.

INTRODUCTION

The photocatalytic degradation of organic pollutants has attracted considerable attention within the past decade as a means of wastewater treatment. Research in this area has begun with the work of Carey et al. [1] who has shown that the irradiation of an aqueous solution of a polychlorinated biphenyl (PCB) in the presence of TiO$_2$ resulted in the disappearance of PCB and the appearance of chloride ions. These intriguing observations have led other researchers to investigate the use of TiO$_2$ and other semiconductor catalysts as a possible means of wastewater treatment [2 - 12].

Detoxification by photocatalysis is attractive for the following reasons: (1) the TiO$_2$ catalyst is cheap, harmless and has a high degree of turnover, (2) there is no need for high photon energy and (3) it can be coated as a thin film on a solid support.

The TiO$_2$ photocatalytic process has been adapted by Nutech Environmental and vigorous research is being conducted in order to make this process a commercially viable technology for water treatment. Using our patented photoreactors (U.S. Patent #4,892,712) the disappearance of various halogenated organic molecules and the appearance of proton (H$^+$) and chloride ions has been demonstrated and the effect of various experimental parameters has been investigated. In this report, we would like to present the results of our investigations on the degradation of mixtures of chlorinated phenols in the presence and absence of electron scavengers.

*Senior author.

H. Al-Ekabi, G. Edwards, W. Holden, A. Safarazadeh-Amiri and J. Story, Nutech Environmental, 511 McCormick Boulevard, London, Ontario N5W 4C8 Canada

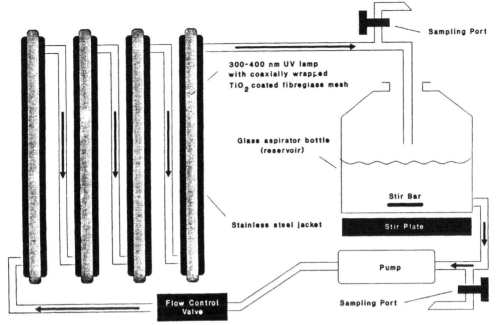

Figure 1: The Nutech Environmental Technology Flow Diagram

EXPERIMENTAL SECTION

The reactor technology is of a simple design (Figure 1). The key part of this technology is the photoreactor which comprises a jacket, a lamp and a photocatalytic sleeve (Figure 2). The lamp emits ultraviolet light in the 300 - 400 nm range and is mounted coaxially within the jacket. Around the lamp lies a sleeve formed of fibreglass mesh that is coated with titanium dioxide (anatase). The anatase is activated by ultraviolet light. Contaminated water flows through the reactor in parallel with the lamp. Typically, 4 - 8 L of contaminated water was circulated through the system for 15 min in the dark. At this point the light was switched on and samples were taken and analyzed on a Hewlett Packard 5890A gas chromatograph using a megabore

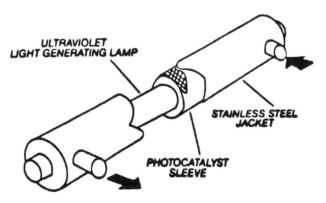

Figure 2 : Schematic of Photoreactor

DB-5 (30 m long, 0.53 mm in diameter) with an electron capture detector operating at 300°C, and nitrogen (14 mL/min) was used as carrier gas.

RESULTS AND DISCUSSION

The degradation of organic pollutants on illuminated TiO_2 is well documented in literature [1 - 12]. Band gap irradiation of TiO_2 promotes an electron from the valence band to the conduction band (e^-_{CB}), generating electron deficiency (or positive hole, h^+_{VB}) in the valence band. The photogenerated "holes" and electrons react with surface adsorbed species, producing ·OH radicals and superoxide ions according to the following equations:

$$TiO_2 + h\nu \longrightarrow e^-_{CB} + h^+_{VB} \qquad (1)$$

$$h^+_{VB} + OH^- (sur.) \longrightarrow ·OH \qquad (2)$$

$$h^+_{VB} + H_2O (ads.) \longrightarrow ·OH + H^+ \qquad (3)$$

$$e^-_{CB} + O_2 (ads.) \longrightarrow O_2^- \qquad (4)$$

The superoxide ion disproportionates to produce hydrogen peroxide.

$$2O_2^- + 2H^+ \longrightarrow H_2O_2 + O_2 \qquad (5)$$

The hydroxyl radical is a strong oxidizing agent and highly reactive and is believed to be responsible for the degradation of organic molecules.

The published work has mostly dealt with mineralization of single organic components. The wastewater is likely contaminated with several organic pollutants and the treatment of organic mixtures is highly desirable. For this reason, we have investigated the degradation of various hydrocarbons such as chlorinated phenols, chlorinated alkenes, chlorinated alkanes and benzene derivatives. The degradation of mixtures of four chlorinated phenols in the presence and absence of electron acceptors will be discussed in this paper.

EFFECT OF ELECTRON ACCEPTORS

A major problem associated with the TiO_2 as a photocatalyst is the electron-hole recombination process. It is very efficient and converts the absorbed photon energy into heat. For instance, the quantum yield of degradation of salicylic acid on illuminated TiO_2 is reported to be 2% [13]. Prevention of electron-hole recombination process is essential for the improvement of the catalytic efficiency of TiO_2.

A strategy to reduce the extent of electron-hole recombination and to increase the photocatalytic efficiency of TiO_2 is to add electron acceptors to the reaction mixture. These reagents trap the conduction band electrons, leaving the "hole" to produce hydroxyl radicals. It is desirable that the electron trapping agent satisfy the following requirements: (1) the electron transfer process should be irreversible, requiring the reagent to decompose following electron trapping, (2) there should be no need for additional waste treatment, (3) if possible, they should produce additional hydroxyl radicals and/or other powerful oxidizing species. We have examined three additives: hydrogen peroxide, ammonium persulphate and oxygen and have found the results very promising.

Effect of Hydrogen Peroxide

Hydrogen peroxide is a suitable electron scavenger and fulfils the requirements indicated above. H_2O_2 can react with conduction band electrons or the superoxide ions [14] generating $\cdot OH$ radicals (Eqs. 6 & 7) which are required for photomineralization.

$$H_2O_2 + e^-_{CB} \longrightarrow \cdot OH + OH^- \tag{6}$$

$$H_2O_2 + O_2^- \longrightarrow \cdot OH + OH^- + O_2 \tag{7}$$

Indeed the addition of H_2O_2 to the reaction mixtures greatly improves the rate of dissapearance of various organic molecules. The degradation of organic mixtures such as chlorinated phenols, chlorinated propenes, and various benzene derivatives are markedly increased in the presence of H_2O_2.

Figure 3 illustrates the degradation of a mixture of four chlorinated phenols such as o-CP, 2,4-DCP, 2,6-DCP and 2,4,6-TCP in the presence and absence of H_2O_2. Clearly the degradation rates are markedly increased when H_2O_2 is added to the solution. Since the lamp used in our reactor emits ultraviolet light in the 300 - 400 nm range with a maximum of 350 nm, and the fact that the extinction coefficient of H_2O_2 at this wavelength is extremely low, we conclude that H_2O_2 increased the degradation rate by acting as an electron acceptor.

Effect of Persulphate

Persulphate ion has been used in the oxidation of various organic compounds [15]. It is an oxidizing agent and is decomposed to sulphate ion and sulphate radical ion by accepting an electron from metal ions such as Fe^{2+} or Cu^+[15]. In principle, it should be able to trap the conduction band electrons or react with the superoxide ions (Eqs. 8 & 9) and accelerate the rate of photocatalytic degradation of organic molecules.

$$S_2O_8^{2-} + e^-_{CB} \longrightarrow SO_4^- + SO_4^{2-} \tag{8}$$

$$S_2O_8^{2-} + O_2^- \longrightarrow O_2 + SO_4^- + SO_4^{2-} \tag{9}$$

Thus by trapping the conduction band electrons, the valence band "hole" is left to produce $\cdot OH$ radicals. In addition, the sulphate radical ions can react either with the OH^- ion to generate $\cdot OH$ radicals (Eq. 10) or it can directly oxidize the organic compounds (Eqs. 11 & 12).

$$SO_4^- + OH^- \longrightarrow SO_4^{2-} + \cdot OH \tag{10}$$

$$(11)$$

$$(12)$$

Indeed, $S_2O_8^{2-}$ significantly improves the rate of disappearance of many organic molecules. For instance, Figure 4 shows the degradation of four chlorinated phenols (o-CP, 2,4-DCP, 2,6-

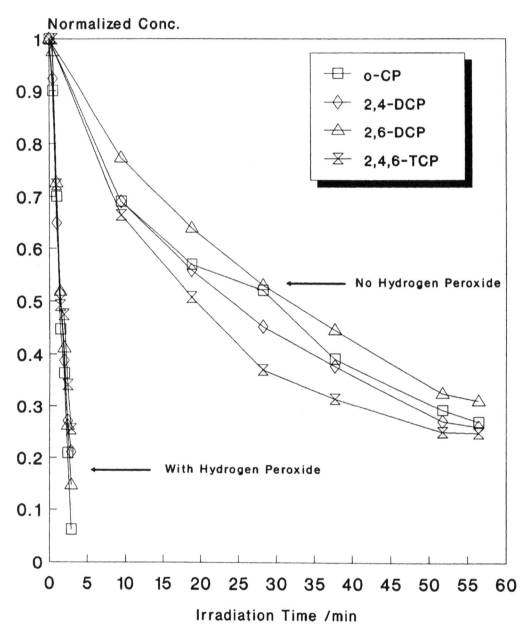

Figure 3 : The photocatalytic degradation of a mixture
containing four chlorinated phenols (40 ppm each) in the
presence and absence of H_2O_2

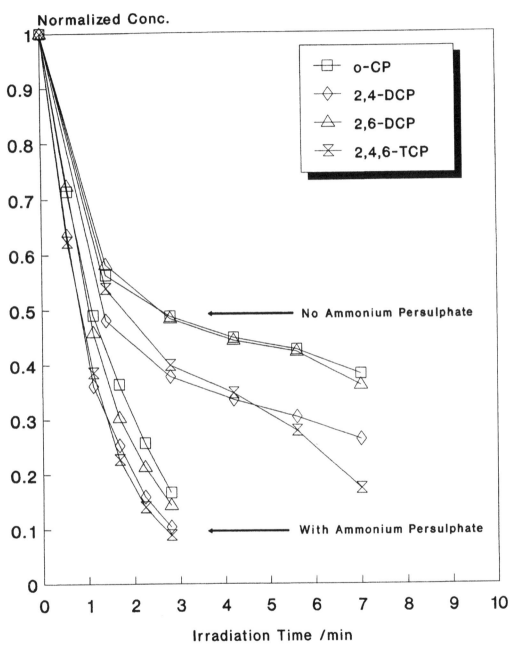

Figure 4 : The photocatalytic degradation of a mixture
containing chlorinated phenols (10 ppm each) in the presence
and absence of $(NH_4)_2S_2O_8$

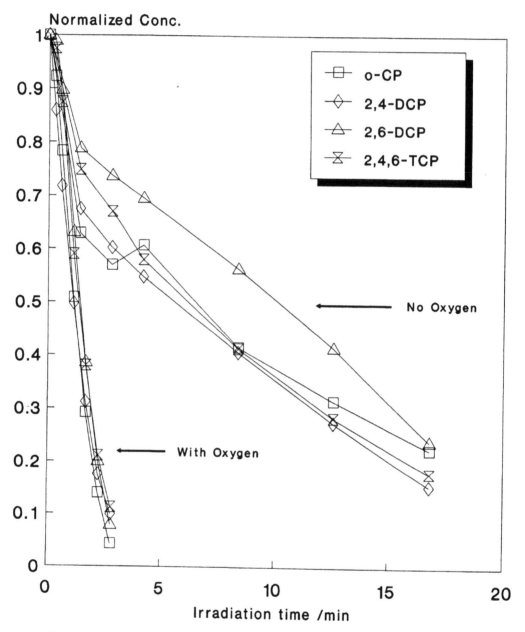

Figure 5 : The photocatalytic degradation of chlorinated
phenols (10 ppm each) with and without added oxygen

DCP and 2,4,6-TCP at 10 ppm each) in the presence and absence of $S_2O_8^{2-}$. Obviously, $S_2O_8^{2-}$ significantly increased the degradation rate of all the components involved in the mixture.

Effect of Oxygen

Oxygen is an essential ingredient of photomineralization and traps the conduction band electrons, generating superoxide ions (Eq. 4). The latter is believed to react with protons, producing hydrogen peroxide (Eq. 5). Oxygen will not only trap the conduction band electrons and prolong the life of the valence band "holes", but it will also produce additional $\cdot$OH radicals (Eqs. 5 - 7). The effect of oxygen on the degradation of a mixture of four chlorinated phenols is shown in Figure 5. Obviously, oxygen enhances significantly the degradation rates of the four components involved in that mixture.

CONCLUSIONS

Our results clearly indicate that the TiO_2 photocatalytic degradation of organic pollutants presents a good opportunity for water treatment. The TiO_2 process will offer a clean, inexpensive and simple technology to remove organic pollutants from water streams.

REFERENCES

1. J.H. Carey, J. Lawrence and H.M. Tosine, *Bull. Environ. Contam. Toxicol., 16,* 697, (**1976**).
2. A.L. Pruden and D.F. Ollis, *J.Catal., 82,* 404, (**1983**).
3. C.-Y. Hsiao, C.-L. Lee and D.F. Ollis, *J. Catal., 82,* 418, (**1983**).
4. H. Al-Ekabi and N. Serpone, *J. Phys. Chem., 92,* 5726, (**1988**).
5. E. Pelizzetti, C. Minero, V. Maurino, A. Sclafani, H. Hidaka and N. Serpone, *Environ. Sci. Technol., 23,* 1380, (**1989**) and references cited therein.
6. R.W. Matthews, *J. Chem. Soc. Faraday Trans. 1., 85,* 1291, (**1989**) and references cited therein.
7. C.S. Turchi and D.F. Ollis, *J. Catal., 119,* 483, (**1989**) and references cited therein. 8. Y.M. Xu, P.E. Menassa and C.H. Langford, *Chemosphere, 17,* 1971, (**1988**).
9. H. Al-Ekabi, N. Serpone, E. Pelizzetti, C. Minero, M.A. Fox and R.B. Draper, *Langmuir, 5,* 250, (**1989**).
10. P. Skala, H. Whittaker, H. Al-Ekabi, M. Robertson, A. Safarzadeh-Amiri and R. Henderson, *Technology Transfer Conference, Ontario Ministry of the Environment, Toronto, Analytical Methods Proceeding,* 171, (**1988**).
11. H. Al-Ekabi, A. Safarzadeh-Amiri, M. Robertson, J. Story and B. Yeo, *Technology Transfer Conference, Ontario Ministry of the Environment, Toronto, Water Quality Research Proceedings,* 379, (**1989**).
12. H. Al-Ekabi, A. Safarzadeh-Amiri, J. Story, W. Holden, and G. Edwards,*Technology Transfer Conference, Ontario Ministry of the Environment, Toronto,Water Quality Research Proceedings,* 275 - 286, (**1990**).
13. R.W. Matthews, *J. Phys. Chem., 91,*3328, (**1987**).
14. K. Tanaka, T. Hisanaga and K. Harada, *New J. Chem. 13,* 5, (**1989**).
15. C.W. Walling, D.M. Camaïoni and S.S. Kim, *J. Amer. Chem. Soc., 100,* 4814, (**1978**).

JESSEMING TSENG
C. P. HUANG

Photocatalytic Oxidation Process for the Treatment of Organic Wastes

ABSTRACT

Upon the irradiation, photocatalyst such as TiO_2, produces electrons and positive holes on the particle surface. The positive hole reacts with hydroxide on the TiO_2 surface to form hydroxyl radicals which can oxidize organic compounds such as chlorinated phenols to Cl^- and CO_2 in the presence of oxygen. Oxygen plays an important role in trapping electrons on the TiO_2 surface and inhibiting electron/hole recombination. Depending upon the degree of chlorination the decomposition rate decreases in the order: monochlorophenol > dichlorophenol > trichlorophenol. With the same degree of chlorination, the Cl atom at para-position tend to be replaced easily by OH radical. Therefore the decomposition rate increases in order: 4CP > 3CP >2CP for monochlorophenols and 26DCP = 25DCP > 34DCP = 24DCP > 23DCP for dichlorophenols. Symmetrical nature of chlorophenols and their intermediates facilitates attack by OH radical due to a favorable orientation. Low yield of carbon dioxide and chloride indicated that photolytic process can not convert chlorophenol into mineral acids.

KEYWORDS

Photocatalytic oxidation, photocatalyst, TiO_2, chlorophenols

INTRODUCTION

The groundwater contamination with organic substances especially chlorophenol compounds has caused great public concern[1,2]. Due to low concentration levels of these organic compounds it is difficult to remove them by conventional treatment processes. Activated carbon adsorption is the most popular process for removing low concentration of organic compounds. Regeneration of activated carbon can be costly[3]. Recently, UV-ozonation process has become a main attraction for removing organic precursors from drinking water. Based on the strong OH absorption bands in the flash photolysis of ozone-water mixture, Norrish and Wayne[4] have proposed that hydroxyl radicals are present in the treatment system. They have also proposed that following the production of oxygen radicals, hydroxyl radicals and hydroperoxyl radicals were formed. However ozone is not stable in solution. Eisenhauer [5,6] has reported the presence of various intermediate products such as catechol and o-quinone during phenol ozonation. This can be a problem for water treatment practice.

As a matter of fact hydrogen peroxide possesses a high reactivity toward organic compounds over a wide pH range without producing toxic by-products. Hydrogen peroxide is a relatively stable oxidation agent compared to ozone. It has been used to decompose inorganic sulfides and thiosulfate[7]. However without the presence

Jesseming Tseng and C. P. Huang, Department of Civil Engineering, University of Delaware, Newark, Delaware, U.S.A.

of catalysts, it is difficult to decompose aromatic compounds such as phenols and chlorinated hydrocarbon. In practice, metal catalysts, e.g. Cu^{2+} or Fe^{2+} have been added to enhance its oxidation capacity. For example, the famous Fenton's reagent was developed by the addition of a soluble iron salt to an acid hydrogen peroxide solution. Barbeni et al.[8] have investigated the degradation of chlorophenol by Fenton's reagent in the presence of oxygen and reported the formation of some aliphatic species as possible intermediates.

Ultraviolet (UV) illumination also is another alternative to degrade organic compounds in dilute solutions[9]. UV light in its own right is not effective in destructing organic chemicals. However, a remarkable result can be observed when it is used together with hydrogen peroxide[10,11] or ozone[12]. The mechanism by which UV-hydrogen peroxide breaks down organic molecules is similar to that for UV-ozone. The ultraviolet light promotes the breakdown of hydrogen peroxide to form hydroxyl radicals. The highly active hydroxyl radicals can react with a wide range of organic compounds. The combination of UV-H_2O_2 or UV-ozone exhibits a reaction rate that is faster than that of hydrogen peroxide or ozone alone. The oxidation is also much more complete than H_2O_2 or O_3 alone yielding final products such as H_2O, CO_2 and low molecular compounds such as acetic acid and oxalic acid. Sundstrom et al.[13-15] demonstrated the destruction of pollutants, benzene and trichloroethylene, by UV-H_2O_2. A pseudo first order reaction was reported. They have also proposed a mechanism which involves hydroxyl radical species. However the UV/H_2O_2 is very selective to the range of UV light. In general a 253.7 nm of wavelength is needed.

Photocatalytic oxidation processes has recently gained great attention in treating organic wastes[16-18]. Semiconductors were selected to generate electron/hole pairs under light illumination. Titanium dioxide, upon illumination with UV light, produces electrons and positive holes. It has been reported the photocatalytic oxidation of phenol and chlorophenols by TiO_2 under ultra violet light illumination in the previous studies[19,20]. Factors such as oxygen, temperature, pH, concentration of the photocatalyst, and concentration of phenol on the decomposition of phenol that may affect the oxidation process were thoroughly examined. Results show that oxygen plays one of the most important roles in the oxidation reaction.

In general photocatalytic processes can be described in Figure 1. The photons strike the particle surface to elevate electrons from the valence band to the conduction band and leave holes in the valence band. The hole and electron can either recombine or diffuse to the surface. Migrated holes can react with species on the surface and perform further reactions. Lack of electron acceptors or hole donors in the liquid phase may cause a surface recombination process to occur.

This study was undertaken to examine the effects of oxygen and light intensity on photocatalytic oxidation. Different chlorophenols were used to study the dechlorination and carbon dioxide production. Comparison of photolytic process with photocatalytic process was also made.

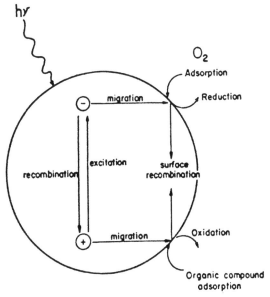

Figure 1.   Schematic Diagram Showing the Generation of Holes/Electrons on Semiconductor and Consequential Interfacial Reactions.

263

METHODS AND MATERIALS

A commercial grade titanium dioxide(R-σ107) was provided by the Du Pont Company. Pretreatment of TiO_2 was described in previous study[19]. The light source used to illuminate TiO_2 samples was a 1,600 watt medium pressure mercury vapor discharge lamp (American Ultraviolet Co.). The spectral irradiance of the UV lamp ranges from 228 to 420 nm and is 260 watts/m^2 at a distance of 1 meter from the light source according to information provided by the manufacturer[21]. A light intensity at 365 nm at 1 meter distance from light source is 17.0 watts/m^2 as detected by a UV-radiometer (Model 365H, Spectronics Co.).

All photocatalytic oxidation experiments were performed in Pyrex tubes containing 15 mL of 10^{-3} M organic in solution and 0.15 g of TiO_2. The experimental set up for the test tube reactors is shown in Figure 2. Test tubes were tightened with teflon septa air-tight caps. The phenol solution was aerated with nitrogen or oxygen from the N_2 prior to experiments. Samples were shaken over a reciprocal shaker (American Optical Co.) with 180 strokes per minute to insure complete mixing. The temperature of the samples was controlled by a thermostat pump in conjunction with a cooler. At the end of a given reaction time, samples were filtered using 0.45 μm microfilters (Gelman, Supor-450, 25mm dia.). All chlorophenols were analyzed by the GC/FID (Hewlett Packard, Model 5890) using HP-1 capillary crosslinked methyl silicone gum column with 0.33 μm film thickness (0.20 mm I.D. x 12.5 m length). The temperature programs were used depending on the property of chlorophenol compounds.

For the gas phase studies, a gas syringe was used to withdraw gas from the test tube. To determine the total quantity of gas, a pressure lok gas syringe was used to withdraw gas mixture from the test tube. The compositions of the gas phase in the head space were detected by a GC/MS (Hewlett Packard, Model 5890/5970) under isothermal conditions. An ion chromatograph program was used to analyze the mass spectrum at 44.0 a.m.u. for CO_2. Chloride was measured by an Orion (Model 96-17B) specific combination chloride electrode.

RESULTS AND DISCUSSIONS

Effect of Oxygen Concentration

As shown in Figure 3, the kinetic results of phenol decomposition with varying dissolved oxygen. Theoretically, one mole of phenol can be completely oxidized with seven moles of oxygen. Oxygen not only initiates the reaction as an electron acceptor but also makes itself for a sequence of oxidation reactions further. As expected the reaction rate increases with increasing concentration of oxygen.

It is evident that the effect of adsorbed oxygen on metal oxide under UV illumination involves the structural alteration of oxide. The exchange between lattice oxygen and adsorbed oxygen was studied by an oxygen isotope exchange technique at room temperature for n-type illuminated metal oxides[22]. An efficient exchange of oxygen between the gas phase and the illuminated TiO_2 surface suggests that the photoadsorption of oxygen initiates a series of subsequent reactions including radical formation reactions and organic oxidations. Several oxygen species such as O^-, O_2^-, O_3^-, O_2^{2-} and O^{2-} are detected by electron spin resonance(e.s.r.)[23]. However amount of the photoadsorbed oxygen is relatively low with a maximum value of 0.3 molecule/nm^2 at a few hundred pascals[24].

Augustine and his co-workers[25] studied the photo-activity of fully hydroxylated TiO_2 in the presence of oxygen. Reducing water molecules and hydroxyl groups on TiO_2 surface by thermal treatment leads to a decrease in oxygen photoadsorption. They reported two different oxygen photoadsorption mechanisms depending on the degree of water adsorbed on the TiO_2 surface. Due to the presence of water molecules, adsorbed oxygen, trapping one free electron, reacts with water molecules resulting in the formation of hydroperoxyl radicals($\cdot HO_2$). If the amount of water molecule on TiO_2 surface is small oxygen radicals, as mentioned above[23], instead of hydroxyl radical and hydroperoxyl radicals are generated.

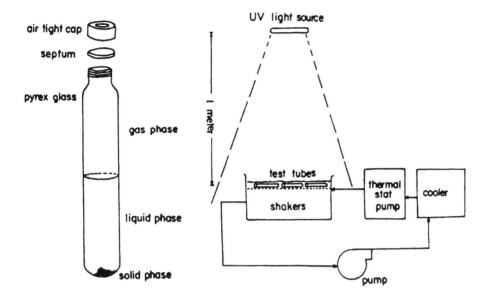

Figure 2. Schematic Diagram of Experimental Set Up for Photocatalytic Oxidation.

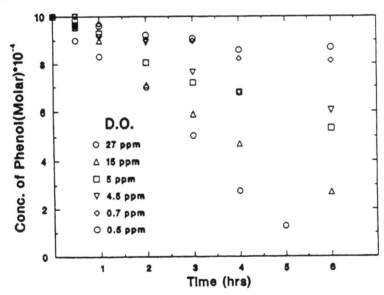

Figure 3.   Effect of Oxygen on Decomposition of Phenol.
Experimental conditions: 10 g/L TiO$_2$, 10^{-3} M phenol,
hν = 17 Watt/m^2, I= 5x10^{-2} M NaClO$_4$, O$_2$ atmosphere, temperature =25°C.

<u>Effect of Light Intensity</u>

In order to excite TiO$_2$ to generate electrons and holes, a wavelength of 387nm or 3.2 eV band gap energy is needed. Increasing light intensity provides larger amounts of photons to the surface of TiO$_2$ to increase the number of electron and hole pairs. Figure 4 shows the result of variation in light intensity on the photocatalytic oxidation. Increases in light intensity enhance the reaction rate. Although ultraviolet light exists only less 1-2%[26] in natural light the following results show the photocatalytic oxidation could be still feasible under natural conditions.

Figure 5 shows the results of phenol decomposition by TiO$_2$ catalyst at various pH values under solar radiation. A minimal effect of pH on phenol removal was found. This implies that the practicality of using TiO$_2$ catalyst for the treatment of toxic wastes or purification of water in aquatic systems. As reported our previous studies[19], the pH effect on phenol decomposition is insignificant due to the mechanism of photocatalytic processes. The photocatalytic process using solar energy as a light source is promising. Although the reaction time is much longer than the UV system in the laboratory, it can be improved by: 1) increasing light intensity via concentrating solar radiation by using a light collector[27], 2) mixing with alumina oxide[28], and 3) addition of hydrogen peroxide[29].

Monochlorophenol

Figure 6 presents the monochlorophenol photodecomposition over TiO$_2$ with different position of chloride under UV illumination. The results show the para-position is much favorable for chlorophenol decomposition. The mechanisms for phenol photocatalytic decomposition suggests that OH radical attacks the C4 (para) position of phenol as a result of hydroquinone formation. It seems that a similar mechanism can be applied

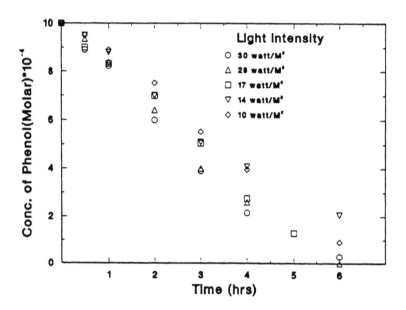

Figure 4.   Effect of Light Intensity on Decomposition of Phenol.
Experimental conditions: 10 g/L TiO$_2$, 10^{-3} M phenol,
I= 5x10^{-2} M NaClO$_4$, O$_2$ atmosphere, temperature =25°C.

266

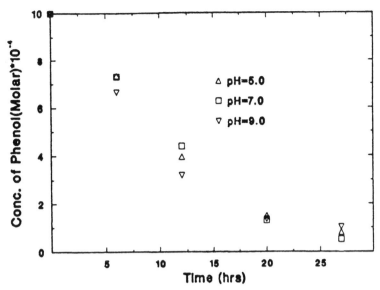

Figure 5.   Effect of pH on Photocatalytic Oxidation under
             Experimental conditions: 10 g/L TiO_2, 10^{-3} M phenol,
             temperature= 16-20 °C, I= 5×10^{-2} M $NaClO_4$.

to chlorophenol decomposition. For 2- and 3CP the attack of OH radical on C4 (para) position leads to the formation of intermediate (e.g. chlorohydroquinone):

The further reaction involves the C-Cl bond scission by OH radicals and hydroxyhydroquinone may be another intermediate before subsequent decompositions.   Competition reaction between chlorophenol and intermediates slows down the chlorophenol photocatalytic decomposition as shown in Figure 6.

267

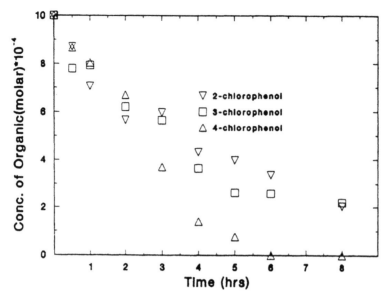

Figure 6.   Photocatalytic Decomposition of Monochlorophenols.
Experimental conditions: 10^{-3} M organic, pH=7.0,
hv = 17 Watt/m^2, O$_2$ atmosphere, I = 5x10^{-2} M NaNO$_3$,
temperature = 25°C, 10 g/L TiO$_2$.

For 4CP the C-Cl bond scission occurs directly via OH radical attack to form a hydroquinone intermediate:

The evidence of chloride production also supports this mechanism(Figure 7). The chloride production rate is faster for 4CP than for 2CP and 3CP.

Direct photolysis is not of interest in this study however it contributes to some extent of decomposition as well. Pichat[30] reported the disappearance of 4CP under illumination at $\lambda > 300$ nm is very fast. However it is very slow at $\lambda > 340$ nm. The wavelength used in the study is greater than 325 nm. Therefore organic decomposition may be influenced by direct photolysis. Table 1 lists the results obtained in different decomposition processes.

It is clear that the direct photolysis process for 2- and 3-chlorophenol leads to a low chloride production while 4CP attains a complete chloride conversion after 8 hours. Boule[31] reported that the direct photolysis is more efficient for 2CP and for 3CP due to the difference in the polarization of the excited molecules. Regardless of the rapid disappearance of 2CP, a complete mineralization is not observed due to the low CO$_2$ yield.

Although there is no direct evidence to predict the final products of photolysis for 2- and 3CP it is likely to form chlorinated biphenyls which are more toxic. The photolysis of 4CP leads to the substitution of the Cl by an OH group because of the polarization of the C-Cl bond in the excited

Table 1. Photodecomposition of Monochlorophenol.

Chlorophenol	Removal* (%)	Cl⁻ produced 10⁻⁴ (M)	CO_2 produced(10^{-5} moles)	
			A	B
Photolysis				
2CP	100	2.71	9.00	1.20
3CP	64	3.31	5.77	0.37
4CP	100	10.0	9.00	1.89
Photocatalytic decomposition				
2CP	80	8.00	7.20	6.43
3CP	78	7.73	7.02	6.77
4CP	100	10.0	9.00	9.20

* @ 8 hours experiments.
A: theoretical; B: experimental

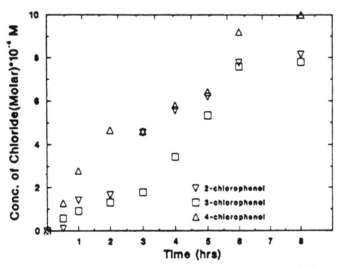

Figure 7.   Chloride Production by Photocatalytic Decomposition of Monochlorophenols.
Experimental conditions: 10^{-3} M organic, pH=7.0,
hν = 17 Watt/m², O_2 atmosphere, I= 5×10^{-2} M $NaNO_3$,
temperature = 25°C, 10 g/L TiO_2.

state[30]. As a result hydroquinone is formed and the release of chloride is expected. However due to the difficulty of replacement of chloride by OH radical to proceed in further decomposition the chlorophenol may not be completely oxidized.

Dichlorophenol

The result of the photocatalytic decomposition of dichlorophenols over TiO_2 is depicted in Figure 8. The disappearance rate of chlorophenol for 25DCP and 26DCP is faster than that of other dichlorophenols. No direct evidences for the formation of intermediates during dichlorophenol photodecomposition. However it is proposed that the OH radical attacks the C4 position of chlorophenol at the onset of the reaction. Apparently the C-Cl bond on the para-position of 24DCP or 34DCP is attacked directly by OH radical resulting the higher chloride production(Figure 9). After releasing the first chloride on the C4 position the intermediate, hydroxyhydroquinone, is formed for 24DCP and 34DCP. Nevertheless further study is necessary to confirm this hypothesis.

Following the attack by OH radical on the C4 position of 25DCP and 26DCP, symmetrical structures of intermediates are formed (e.g. 2,5-dichloro-hydroquinone and 2,6-dichloro-hydroquinone). Due to the symmetry of chemical structure the replacement of chloride by OH radicals is expected to be much easier. Evidently, the result of 23DCP photodecomposition shows that 23DCP is relatively stable with low chloride production. Table 2 summarizes the results between photolytic and photocatalytic dichlorophenol decomposition.

Table 2. Photodecomposition of Dichlorophenol Decomposition in Different Processes.

Chlorophenol	Removal* (%)	Cl$^-$ produced 10^{-3} (M)	CO$_2$ produced(10^{-5} moles) A	B
Photolysis				
2,3DCP	57	0.70	5.13	1.20
2,4DCP	78	0.94	7.02	0.37
2,5DCP	-	0.81	-	1.89
2,6DCP	-	2.71	-	1.20
Photocatalytic decomposition				
2,3DCP	70	0.82	6.30	3.35
2,4DCP	77	1.30	6.93	7.88
2,5DCP	96	1.70	8.64	8.45
2,6DCP	91	2.00	8.19	8.00
3,4DCP	80	1.53	7.20	6.97

* @ 8 hours experiments.
A: theoretical: B: experimental

Trichlorophenol

As shown in Figure 10, the photocatalytic decomposition of trichlorophenol is demonstrated at pH 7.0. The disappearance of trichlorophenol shows a fair consistence for individual trichlorophenol by HPLC analysis. According to the result of chloride production in Figure 11, it shows that dechlorination process are more complete for 245TCP than for 235TCP under photocatalytic decomposition. This suggests that the chloride at C4 position is replaced by OH radical therefore high chloride concentration is observed. In order to verify the completeness of the photocatalytic decomposition process, the production of carbon dioxide was monitored and is presented in Figure 12. It indicates that 245TCP has higher carbon dioxide conversion than 235TCP under photocatalytic decomposition. Table 3 lists the results between photolytic and photocatalytic decomposition for trichlorophenol.

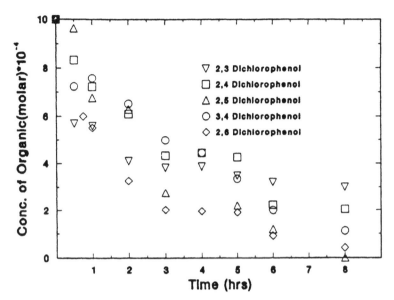

Figure 8. Photocatalytic Decomposition of Dichlorophenols.
Experimental conditions: 10^{-3} M organic, pH=7.0,
$h\nu = 17$ Watt/m^2, O$_2$ atmosphere, I= 5x10^{-2} M NaNO$_3$,
temperature=25°C, 10 g/L TiO$_2$.

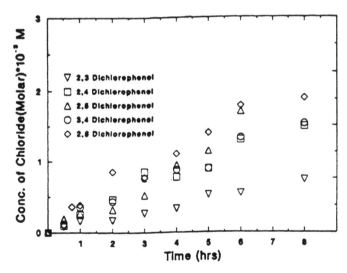

Figure 9. Chloride Production by Photocatalytic Decomposition of Dichlorophenols.
Experimental conditions: 10^{-3} M organic, pH=7.0,
$h\nu = 17$ Watt/m^2, O$_2$ atmosphere, I= 5x10^{-2} M NaNO$_3$,
temperature= 25°C, 10 g/L TiO$_2$.

Apparently the decomposition of 235TCP by photocatalytic processes produces chlorinated intermediates in the early stage. It is proposed that the hydroxyl radical attacks the C4 position to form chlorohydroquinone as follows:

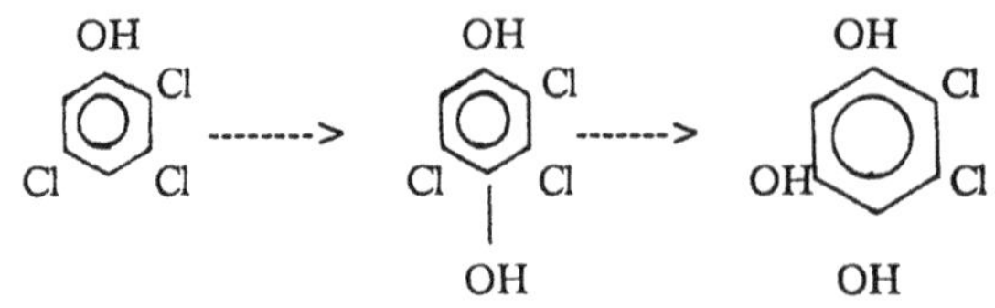

The formation of 2,3 chloro-hydroxyhydroquinone causes the low conversion of carbon dioxide and chloride production as well. While in the case of 245TCP, the Cl atom at C4 position is replaced by OH to form symmetrical 2,5 chloro-hydroquinone. Either one of chlorides can be replaced by an OH again:

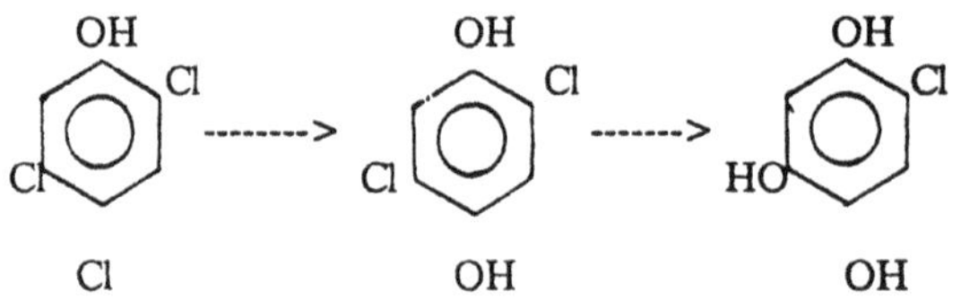

Table 3. Photodecomposition of Trichlorophenol.

Chlorophenol	Removal[*] (%)	Cl⁻ produced 10^{-3} (M)	CO_2 produced(10^{-5} moles)	
			A	B
Photolysis				
235TCP	58	1.37	5.22	0.69
245TCP	86	2.57	7.74	0.61
246TCP	100	0.05	9.00	-
Photocatalytic decomposition				
235TCP	79	1.22	7.11	3.66
245TCP	90	2.43	8.11	7.20
246TCP	87	2.51	7.83	-

[*] @ 8 hours experiments.
A: theoretical; B: experimental

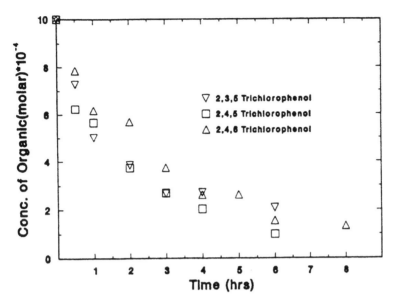

Figure 10.   Photocatalytic Decomposition of Trichlorophenols.
Experimental conditions: 10^{-3} M organic, pH=7.0,
hv = 17 Watt/m^2, O$_2$ atmosphere, I= 5x10^{-2} M NaNO$_3$,
temperature= 25°C, 10 g/L TiO$_2$.

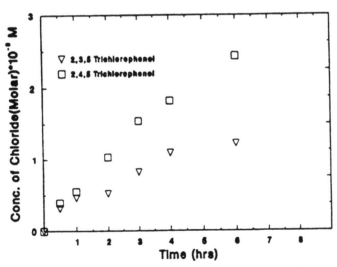

Figure 11.   Chloride Production by Photocatalytic Decomposition of Trichlorophenols.
Experimental conditions: 10^{-3} M organic, pH=7.0,
hv = 17 Watt/m^2, O$_2$ atmosphere, I= 5x10^{-2} M NaNO$_3$, temperature= 25°C, 10 g/L
TiO$_2$.

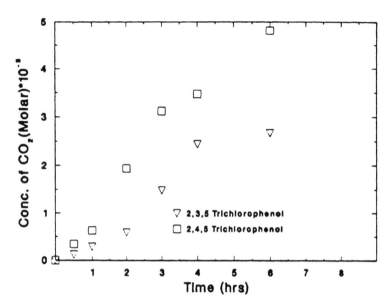

Figure 12.   Carbon Dioxide Yield by Photocatalytic Decomposition of Trichlorophenols. Experimental conditions: 10^{-3} M organic, pH=7.0, hv = 17 Watt/m^2, O_2 atmosphere, I= 5×10^{-2} M NaNO$_3$, temperature= 25°C, 10 g/L TiO$_2$.

CONCLUSIONS

Photocatalytic oxidation of phenols is affected by oxygen concentration and light intensity. Hydroxyl radicals, highly active radicals, are probably responsible for the initiation of phenol decomposition reaction. The formation of OH radicals can be ascribed to (1) photoproduced holes reacting with hydroxyl groups on TiO$_2$; and (2) oxygen trapping photoproduced electrons and then reacting with hydroxyl group. At neutral pH both reactions can occur simultaneously. Increasing light intensity increases the rate of phenol oxidation.

Hydroxyl radicals appear to be the most important species for photocatalytic decomposition. It is postulated that the hydroxyl radicals attack the C4 position of chlorophenols and form hydroxy-hydroquinone or chlorohydroquinone intermediates. Depending upon the degree of chlorination the decomposition rate decreases in the order: monophenols > dichlorophenols > trichlorophenols. With same degree of chlorination, the Cl atom at para-position tend to be replaced easily by OH radicals. Therefore the decomposition rate increases in the following order: 4CP > 3CP >2CP for monochlorophenol; 26DCP=25DCP > 34DCP = 24DCP > 23DCP for dichlorophenols, and 245TCP > 235TCP for trichlorophenols . Symmetrical nature of chlorophenols and their intermediates ehances attack by the OH radical due to a favorable orientation.

In general the toxicity of chlorinated aromatic is much stronger than nonchlorinated aromatic in the environment. Although the mechanism of photocatalytic decomposition of trichlorophenol remains unclear the dechlorination process may convert chlorinated phenol to compounds which can be degraded in nature or with existing biological treatment processes.

Photocatalytic oxidation can be an effective process in degrading hazardous organic substances such as phenols. Toxic organics can be broken into an environmentally acceptable chemicals such as CO$_2$. TiO$_2$ is an inexpensive photocatalyst[32] and should be affordable to many water and wastewater treatment utilities. Moreover, TiO$_2$ is stable upon photo-irradiation compared to ZnO, ZnS, and CdS.

ACKNOWLEDGMENT

The research on which this paper is based was supported in part by the United State Environmental Protection Agency research grant No. R-815081. This paper has not been reviewed by the funding agency. Any mentions of chemicals and process do not necessary imply the endorsement of the funding agency.

REFERENCES

1. Westrick, J. J., Mello, J. W. and Thomas, R. F. 1984."The Groundwater Supply Survey", J. Am. Water Works Assoc., 76, No.5, pp52-59

2. Folkard. G. K. 1986. "The Significance, Occurrence and Removal of Volatile Chlorinated-Hydrocarbon Solvents in Groundwater", Water Pol. C., Vol. 55. No.1, pp63-70

3. Singer, P. C. and Yen, C. Y. 1981. " Adsorption of Alkylphenols by Activated Carbon: Chapter 8", in Activated Carbon Adsorption, vol.1, Edited by Suffet, I. H. and McGuire, M. J., Ann Arbor Science Publishers, pp167-189

4. Norrish, R. G. W. and Wayne, R. P. 1965. "The Photolysis of Ozone by Ultraviolet Radiation II. The Photolysis of Ozone Mixed with Certain Hydrogen-containing Substances". Proc. Roy. Soc. Ser. A288, pp361-370

5. Eisenhauer, H. R., 1968 "The Ozonation of Phenolic Waste", J. Water Poll. Cont. Fed., 40, pp 1887-1899

6. Eisenhauer, H. R., 1964 "Oxidation of Phenolic The Ozonation of Phenolic Waste", J. Water Poll. Cont. Fed., 36, pp 1116-1128

7. Bull, R. A. and McManamon, J. T. 1990. "Supported Catalysts in Hazardous Waste Treatment: Destruction of Inorganic Pollutants in Wastewater with Hydrogen Peroxide", in Emerging Technologies in Hazardous Waste Management, edited by Tedder, D. W. and Pohland, F. G., ACS Sym. Ser. A422, pp52-66

8. Barbeni, M., B., Minero, C., Pelizzetti, E., Borgarello, E. and Serpone, N. 1987. "Chemical Degradation of Chlorophenols with Fenton's Reagent (Fe^{2+} + H_2O_2)", Atmosphere, 10(12), pp2225-2237

9. Zeff, J. D. 1977. "UV-OX Process for the Effective Removal of Organics in Wastewaters", AIChE Symp. Ser. 73, 167, p206

10. Koubek, E., 1975. "Photochemically Induced Oxidation of Refractory Organics with Hydrogen Peroxide, Ind. Eng. Chem. Process Des. Dev. 14, No.3 pp358-350

11. Ho, P. C. 1986. "Photooxidation of 2,4-dinitrotoluene in Aqueous Solution in the presence of Hydrogen Peroxide", Environ. Sci. Technol., 20, p260

12. Peyton, G. R., Huang, Burleson, J. L. and Glaze, W. H. 1982. "Destruction of pollutants in Water with Ozone in Combination with UV Radiation. 1. General Principles and Oxidation of Tetrachloroethylene", Environ. Sci. Technol., 16, pp448-453

13. Sundstorm, D. W., Klei, H. E., Nalette, T. A., Reidy, D. J. and Weir, B. A. 1986. "Destruction of Halogenated Aliphatics by Ultraviolet Catalyzed Oxidation with Hydrogen Peroxide", Hazard. Waste Hazard. Mat., 3, p101

14. Sundstorm, D. W., Weir, B. A. and Klei, H. E. 1989. "Destruction of Aromatic Pollutants by

Ultraviolet Catalyzed Oxidation with Hydrogen Peroxide", <u>Environ. Prog.</u> 8, p6

15. Sundstorm, D. W., Weir, B. A. and Redig, K. A. 1990. "Destruction of Mixtures of Pollutants by UV-Catalyzed Oxidation with Hydrogen Peroxide", in <u>Emerging Technologies in Hazardous Waste Management</u>, edited by Tedder, D. W. and Pohland, F. G., ACS Sym. Ser A422, pp67-76

16. Al-Ekabi, H. and Serpone, N. 1988. "Kinetic Studies in Heterogeneous Photocatalysis. 1. Photocatalytic Degradation of Chlorinated Phenols in Aerated Aqueous Solutions over TiO_2 Supported on a Glass Matrix", <u>J. Phys. Chem.</u> Vol.92, No. 20, pp5726-5731

17. Pelizzetti, E., Borgarello, M., Minero, C., Pramauro, E., Borgarello, E. and Serpone, N. 1988. "Photocatalytic Degradation of Polychlorinated Dioxins and Polychlorinated Biphenyls in Aqueous Suspensions of Semiconductors Irradiated with Simulated Solar Light", <u>Chemospere</u>, Vol.17, No.3, pp499-510

18. Peral, J., Casado, J. and Domenech, J., "Light-induced Oxidation of Phenol over ZnO Power", 1988. <u>J. Photochem. & Photobio.</u>, 44, pp209-217

19. Tseng, J. M. and Huang, C. P. 1990."Mechanistic Aspects of the Photocatalytic Oxidation of Phenol in Aqueous Solutions, in <u>Emerging Technologies in Hazardous Waste Management: chapter 2.</u> edited by D. W. Tedder and F. G. Pohland, ACS Symposium Series 422, pp12-39

20. J. M., Tseng and C.P. Huang, 1991. "Removal of Chlorophenols from Water by Photocatalytic Oxidation.", <u>Wat. Sci. Tech.</u>, vol 23, pp377-387

21. American Ultraviolet Co. 1987., Technical data for porta-cure irradiators., N. J.

22. Courbon, H., Formenti, M. and Pichat, P. 1977. "Study of Oxygen isotopic Exchange over Ultraviolet Irradiated Anatase Samples and Comparison with the Photooxidation of Isobutane into Aceton", <u>J. Phys. Chem.</u>, 81, pp550-554

23. Andreev, N. S. and Prudnikov, I. M. 1974. "Spectral Investigation of Photosorption of Oxygen and Methane and Photoinitated ESR Signals on Zinc Oxide Mechanisms of Methane Photosorption", <u>J. Kinetics and Catalvtsis</u>, 15, pp636-641

24. Courbon, H., Herrmann, J. M. and Pichat, P. 1984. "Effect of Platinum Deposits on Oxygen Adsorption and Oxygen Isotope Exchange Over Variously Pretreated, Ultraviolet- Illuminated Powder TiO_2", <u>J. Phys. Chem.</u>, 88, pp5210-5214

25. Augustine, R. G., Munuera, G., and Soria, J. 1979. "Photoadsorption and Photodesorption of Oxygen on Highly Hydroxylated TiO_2 Surfaces: Part.2- Study of Radical Intermediates by Electron Paramagnetic Resonance", <u>J. Chem. Soc. Faraday Trans. 1</u>, 75, pp748-761

26. AMETEK, Inc. 1984. <u>Solar Energy Handbook- Theory and Applications</u>, 2nd Ed., Chilton Book Co., p17

27. Pacheo, J. E. and Holmes, J. T., 1990. "Falling- Films and Glass- Tube Solar Photocatalytic Reactors for Treating Contaminated Water", in <u>Emerging Technologies in Hazardous Waste Management</u>, edited by Tedder, D. W. and Pohland, F. G., ACS Sym. Ser. A422, pp77-99

28. J. M., Tseng and C. P. Huang, 1991. "Phenol Decomposition by a TiO_2/Al_2O_3 Photocatalyst under Ultra-violet Light Illumination.", AICHE, National Meeting, San Diego, Aug 21-24,1990

29. Tseng, J. M. 1991. Photocatalytic Oxidation of Phenols by Titanium Dioxide Suspension, Ph. D.

276

Dissertation, University of Delaware, Newark, DE

30.   Pichat, P., 1987. "Powder Photocatalysts Chracterization by Isotopic Exchanges and Photoconductivity; Potentialities for Metal Recovery, Catalyst Preparation and Water Pollutant Removal", in Photocatalysis and Environment- Trends and Applications, edited by M. Schiavello, Kluwer Academic Publishers, The Netherlands, pp 399-424

31.   Boule, P., Guyon, C., Tissot, A., Lemaire, J. 1987."Photochemistry of Environmental Aquatic Systems", ACS Symp. Ser. no.327, pp10-26

32.   Key Chemicals & Polymers, 7ed. 1985., Chemical & Engineering News, American Chemical Society, Washington D. C. p43

Supercritical Water Oxidation

ABSTRACT

Supercritical water oxidation is an innovative, relatively low-temperature process that can give high destruction efficiencies for a wide variety of hazardous chemical wastes. It takes place in a water medium with added oxidant above the critical point of water. The scientific basis for development of supercritical water oxidation is reviewed, with particular emphasis on work at the Los Alamos National Laboratory.

INTRODUCTION

Destructive oxidation of hazardous wastes to carbon dioxide, water, and other small molecules can effectively minimize waste volume and detoxify many hazardous compounds. Incineration in air at atmospheric pressure is the most common oxidation technique currently practiced. Supercritical water is a unique solvent medium in which oxidation can take place at lower temperatures than those of incineration. A supercritical water oxidation system can handle aqueous streams containing organics in relatively low concentrations and offers inherent control over emissions and coupling to energy recovery systems. Supercritical water oxidation takes place in a water medium with added oxidant at temperatures and pressures above the critical point of water (374°C and 22.13 MPa; Figure 1). The properties of supercritical water that have led to its consideration for this process are its gas-like transport properties, densities high enough for reasonable throughputs, and solubility properties. The viscosity of water ranges from about 50 µPa.s near the critical point to about 30 µPa.s under most reported supercritical processing conditions, in comparison to a viscosity of 891 µPa.s at atmospheric pressure and 25°C [1]. Its density ranges from 0.3 g/cm^3 at the critical point to about 0.1 g/cm^3 under processing conditions [1]. High solubilities, with large regions of complete miscibility, have been observed for non-polar organics in supercritical water, including

Cheryl K. Rofer, Earth and Environmental Sciences Division, Los Alamos National Laboratory, P.O. Box 1663, MS D462, Los Alamos, New Mexico, U.S.A.

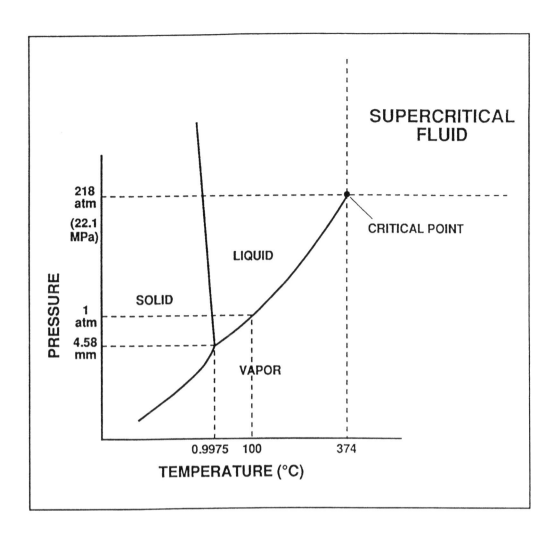

Figure 1. Phase diagram for water. Not drawn to scale.

saturated, [2] unsaturated, [3] aromatic, [4] and halogenated [5] hydrocarbons. Solubilities of nonpolar gases, including oxygen, [6] nitrogen, [7] and hydrogen [8] are also high. Conversely, salts become less soluble in supercritical water than in liquid water, [9] so that they may be separated as part of the process. Although multicomponent phase diagrams in the supercritical region are difficult to predict, the two-component systems suggest that the water, the oxidant, and the compound to be destroyed are likely to form a single phase under typical processing conditions. Within this single phase, chemical kinetics, rather than transport processes, should determine the rate of reaction. In multi-phase systems, concentration gradients across phase boundaries also limit the extent to which compounds can be completely destroyed.

Few studies of reactions in supercritical water have been available in the open literature [10]. High destruction efficiencies have been reported for a number of compounds, [11] and some data on products and global kinetics are available for oxidation of simple compounds in supercritical water [12]. Non-oxidative conversions of coal [13] and biomass [14] in supercritical water have also been studied. Some of the reactions observed in these conversion processes may also occur during the oxidation of complex organics in supercritical water.

A significant patent literature on supercritical water also exists, but review of that literature is beyond the scope of this paper.

Oxidation in supercritical water takes place at about 500 to 800°C in a completely contained system. Thermodynamic considerations predict that almost no oxides of nitrogen should be produced at these temperatures. The effluent can be completely controlled because the system is contained. In addition, because complete mixing is possible in a single-phase supercritical region, reaction kinetics are not diffusion limited. Supercritical water oxidation has the promise of destroying, at reasonable cost, low concentrations of hazardous chemicals in water, such as groundwater contaminated with chlorinated hydrocarbons. The oxidant may be air, O_2, hydrogen peroxide, or another suitable compound. The reactions are exothermic, and depending on the nature and concentration of the hazardous chemicals and the design of the process, excess heat might be used for process heat or to cogenerate electricity.

Many hazardous compounds can be destroyed by oxidation in supercritical water. In principle, any organic compound can be completely oxidized to relatively innocuous compounds. Oxidation of carbon gives carbon dioxide, hydrogen gives water, nitrogen is reported to give ammonia and dinitrogen, phosphorus gives phosphoric acid, sulfur gives sulfuric acid, and the halogens give the corresponding halogen acids. Mineral acid products such as sulfuric, hydrochloric, and phosphoric acids can be neutralized to salts by use of caustics. Supercritical water oxidation appears to be applicable to the destruction of most organic chemicals. Some of the many waste streams that might be treated by supercritical water oxidation include:

> liquid aqueous and organic streams, including machining wastes, paint wastes, automobile grease and lubricant wastes, PCB-contaminated oil, and waste solvents;
> groundwater contaminated with organics;
> stored wastes, including those that contain sludges or other solids;
> soil contaminated by spills or burial of organics;

vermiculite and other mineral absorbers used to clean up spills of organics;

mixed wastes containing both organics and radioisotopes, including uranium machining wastes; and

spent carbon adsorbent.

Because water is the reaction medium, the process can be used for a variety of organic wastes containing water or for water contaminated with organic compounds. There appear to be no technological upper or lower limits to the concentration of organic in water. An economic constraint, however, on the organic concentration arises from the amount of heat generated by oxidation of the organic. The optimum concentration of organic in water will vary with the heat of oxidation of the particular organic compounds present and the engineering design of the apparatus. An engineering tradeoff to be considered in the design of a plant is the organic concentration that generates enough heat to maintain the reaction, but not more than can readily be removed from the processing vessel (autogenic operation). Pure or highly concentrated organic wastes can be diluted with water, whereas fuel or other organic wastes can be added to wastes in low concentrations. Other factors that influence the engineering design include the residence time in the reactor (determined by the chemical kinetics of oxidation of the waste), the physical state of the waste and its oxidation products, and the amounts of waste to be processed.

A great range of scale appears to be possible for supercritical water oxidation plants. Standard pressure-vessel technology can be used to provide small to medium-sized surface installations, which could be mobile or permanently installed to process laboratory or manufacturing wastes. In addition, plants with very large capacities have been proposed, in which a cylindrical heat exchanger and reaction vessel is emplaced in the ground by use of oil-field drilling and well-completion technology [15]. The depth of the cylinder provides some pressure by hydrostatic head, and the emplacement in the earth provides structural strength for containment of the pressure. A waste treatment plant based on oxidation in supercritical water would not require a large land area. Modular plants processing tens of gallons per minute could be skid-mounted and carried on one or two semi trailers. An in-ground unit processing wastewater from a medium-sized city would require about an acre of land. Both the surface and the in-ground plants can be engineered to produce hot, pressurized water, which can provide electrical energy, shaft power, or process steam by means of proven power plant technology.

The Los Alamos National Laboratory is investigating supercritical water oxidation for both the Department of Energy (DOE) Office of Technology Development and the Air Force Engineering and Services Center. In the DOE program, the objectives are to obtain information necessary for evaluation of the technology and for process development, scaling up, and application to Department of Energy wastes. The oxidation of one carbon model compounds in supercritical water has been investigated to determine kinetics and mechanism.[16] These investigations, summarized in this paper, show that oxidation in supercritical water is substantially similar to gas-phase oxidation if the operating pressures and water densities of supercritical water oxidation are properly taken into account.

The objective of the program sponsored by the Air Force is to determine the feasibility of destruction of propellant components in supercritical water. Open burning and open detonation of propellants and their wastes are becoming environmentally unacceptable, and alternative destruction technologies are needed. In this program, several propellant components, including ammonium perchlorate (AP), [17] trinitrotoluene (TNT), and unsymmetrical dimethylhydrazine (UDMH) are being investigated for safe introduction into the supercritical water environment and to determine the products of their destruction in supercritical water. These results are also summarized in this paper.

CHEMISTRY OF THE PROCESS

Because the oxidation appears to take place in a one-phase region, the kinetics of the process may well be controlled by chemical kinetics, rather than by mass transfer. The evidence for reaction in a one-phase region is not conclusive; as stated above, phase diagrams are not available for most systems of interest. However, the residence times for supercritical water oxidation are reported to be seconds to minutes, [18] about an order of magnitude faster than the reported residence times for subcritical wet oxidation [19]. In addition, reaction of organics is reported to be more complete at the shorter residence times for supercritical water oxidation [18].

These two observations together suggest strongly that the kinetics of supercritical water oxidation will be dominated by chemical, rather than mass-transfer, kinetics. Further, the single phase in which reaction occurs is a dense gas. Thus, a sound basis for investigating the chemical kinetics may be found in gas phase reactions. An extensive literature is available on gas-phase elementary reactions, particularly for combustion and atmospheric oxidation. Computational resources are readily available for calculation of overall chemical kinetics from detailed mechanisms composed of elementary reactions. Such mechanisms, once validated, can be used to extrapolate kinetics in ways that are beyond the application of standard global reaction rate expressions. Development of such a capability is one of the objectives of the Los Alamos work. Because waste streams can be highly variable, and because chemical kinetics can vary over orders of magnitude, it is desirable to investigate the probable variation in the kinetics of supercritical water oxidation (and consequently residence times and reactor sizes) as a means of evaluating the usefulness of this technology for application to a wide variety of wastes. The effects of the pressure and high concentrations of water must be properly taken into account in any mechanism derived from the available elementary reaction kinetics data for combustion or atmospheric oxidation.

Methane and its partial oxidation product carbon monoxide were selected as model compounds. Methane was chosen as a model compound because a large data base is available on its elementary reactions. The oxidation mechanism for carbon monoxide is a subset of the methane oxidation mechanism. Experimental kinetics results were available for carbon monoxide at the inception of the project [12b,c]. As part of the DOE-sponsored project, experimental data were obtained for the rates of methane oxidation in supercritical water in a flow reactor. Global rate expressions were derived from these data, and the product distribution and kinetic data were used to validate the elementary reaction mechanisms [16]. In validating the mechanism against experiment, two measures were used: the amount of reaction, indicated by the percent carbon monoxide oxidized, and the product distribution, indicated by the ratio of hydrogen to carbon dioxide concentrations.

A high-pressure low-temperature methane oxidation mechanism developed at Los Alamos was used as the basis for development of a mechanism. The subset of reactions representing carbon monoxide oxidation from from this mechanism was used for development of the mechanism for oxidation of carbon monoxide in supercritical water. Modifications to this mechanism included updating of the rate constants, calculation of rate constants otherwise unavailable from the literature, modification of collision efficiencies for very efficient energy transfer, high-pressure limits for association and dissociation reactions, and the addition of about a dozen reactions of water that are usually excluded from combustion mechanisms because of the low concentrations of water in conventional combustion. The results of simulation with this mechanism in comparison to the experimental data for carbon monoxide oxidation in supercritical water (at 24.6 MPa) are shown in Figure 2. The disagreement with the experimental data appears primarily to result from phenomena such as hydrogen bonding or clustering that may cause nonideal behavior as the critical point of water is approached.

The oxidation of methane takes place at higher temperatures than that of carbon monoxide and therefore should be less affected by the nonidealities occurring near the critical point. The temperature range for methane experiments is 640-703°C at 24.5 MPa, and the water density is 7% to 10% over ideal. For the methane oxidation mechanism, the carbon monoxide mechanism as developed above was substituted into the Los Alamos high-pressure, low-temperature methane oxidation mechanism. No changes were made beyond those changes to the carbon monoxide oxidation mechanism listed above. A comparison of experimental data and simulation results is shown in Figure 3. Both the rate of oxidation and the product ratio show outstanding agreement. The small set of experimental points exhibiting a high slope is from a single experiment and may share a systematic source of error. The calculated results in Figures 2 and 3 differ slightly from those presented earlier [16]. These differences arise from use of improved kinetic constants for some elementary reactions.

A mechanism for acetone oxidation in supercritical water based on the one-carbon mechanisms has been developed, and preliminary data have been obtained for acetone oxidation in a batch reactor. Early results from simulations with this mechanism agree well with the experimental results.

The agreement of a gas-phase elementary reaction mechanism that has been modified for high pressure and high concentrations of water with experimental data for carbon

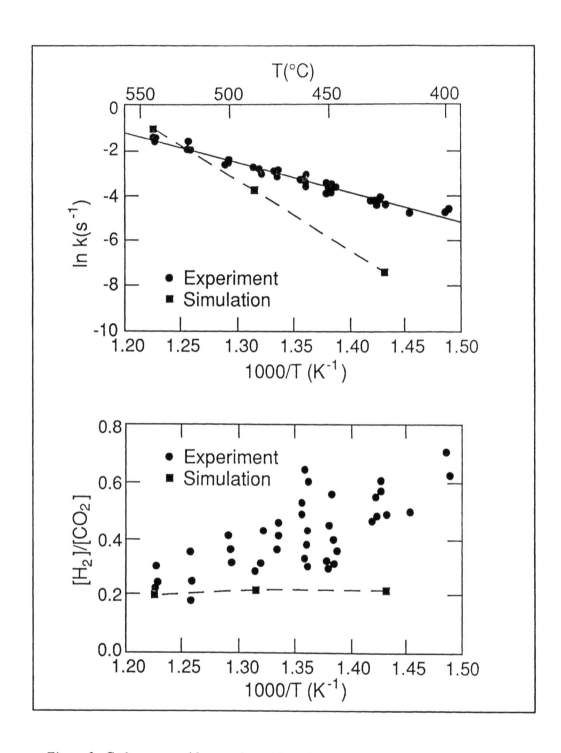

Figure 2. Carbon monoxide experimental results and simulation by Los Alamos supercritical water oxidation mechanism.

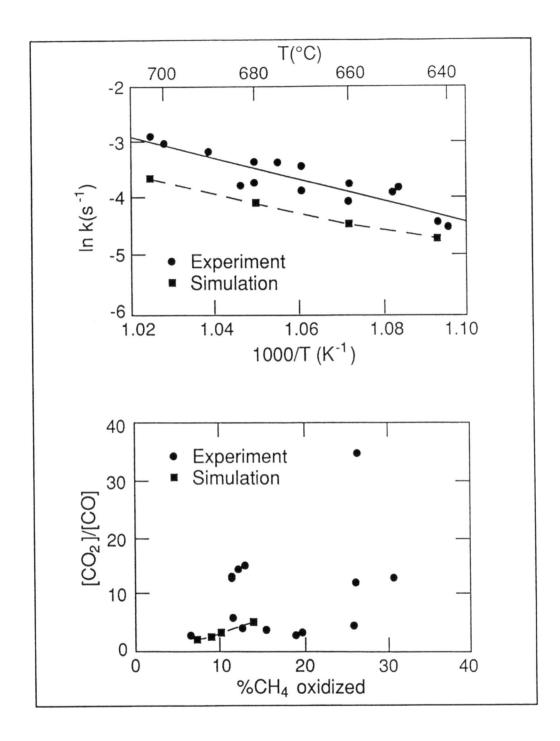

Figure 3. Methane experimental results and simulation by Los Alamos supercritical water oxidation.

monoxide, methane, and acetone from flow and batch reactors is strong support for the hypothesis stated above that the chemistry of supercritical water oxidation is related to combustion and atmospheric oxidation. The reaction mechanism consists of free-radical reactions, and the primary chain carriers are peroxy radicals, as is to be expected in this temperature range for reactions at atmospheric pressure without the large concentrations of water. Additional processes appear to become important near the critical point, some of which may be related to interactions of water molecules in clusters with the reactants [20].

Catalysis of the reactions has not been considered thus far in the calculations discussed above. The agreement of calculation and experiment supports neglecting catalysis. However, low-temperature gas-phase oxidation at atmospheric pressure is known to be extremely sensitive to the condition of the surface of the vessel in which the reaction is carried out [21]. In addition, catalytic amounts of metals may be leached from the metal reaction vessels used for supercritical water oxidation by the water-reactant mixture. These dissolved components could also act as catalysts. With the limited data available, it is not possible to assess potential catalytic effects. The strongest evidence against catalytic effects in these experiments is the agreement of the mechanism with experimental results from both batch (low surface-to-volume ratio) and flow reactors.

For compounds containing elements other than C, H, and O, other types of reaction beyond the basic free-radical mechanism may become important. For example, water may eliminate chloride from chlorinated hydrocarbons via nucleophilic substitution reactions. Other solution-related reactions may take place, particularly in solvated clusters that may form near the critical point.

Ammonium perchlorate and nitromethane have been destroyed, without addition of oxidant or fuel, in a supercritical water flow reactor [17]. Results from these experiments are summarized in Tables I and II. No pressure or temperature excursions were observed. Chlorine (Cl_2), nitrosyl chloride (NOCl), and nitrogen dioxide (NO_2) were not observed in the mass spectrometric analysis of the product gases from ammonium perchlorate, but nitrous oxide (N_2O) and oxygen (O_2) were. The absence of pressure and temperature excursions and the low production of higher nitrogen oxides indicate that supercritical water destruction methods may be suitable for at least some propellants and explosives. These data are insufficient for conclusions about chemical mechanisms. It would not be surprising if some of the reactions of nitrogen-containing species took place by solution-related mechanisms.

SUMMARY

The chemistry of oxidation in supercritical water is being investigated and developed for a usable waste destruction process. This relatively low temperature oxidation has been proposed as a means of destroying hazardous organic wastes, with full control of all effluents. Experimental data and modeling of the oxidation of one-carbon compounds in supercritical water strongly support the hypothesis that oxidation in supercritical water

TABLE I - PRELIMINARY RESULTS FOR AMMONIUM PERCHLORATE

- 0.1 M NH_4ClO_4 (12 grams/liter) + NaOH
- 39 MPa

Temperature:	400°C		
Residence Time (sec)	45	90	180
% NH_4 Recovered	93	80	70
% ClO_4 Recovered	100	100	100

Temperature:	450°C		
Residence Time (sec)	19	39	78
% NH_4 Recovered	27	<0.1	<0.1
% ClO_4 Recovered	60	33	21

Temperature:	475°C		
Residence Time (sec)	16	32	64
% NH_4 Recovered	<0.1	<0.1	<0.1
% ClO_4 Recovered	14	3	0.1

Temperature:	500°C		
Residence Time (sec)	15	30	60
% NH_4 Recovered	<0.1	<0.1	<0.1
% ClO_4 Recovered	<0.05	<0.05	<0.05

TABLE II - PRELIMINARY RESULTS FOR NITROMETHANE

- 0.16 M CH_3NO_2 (10 grams/liter)
- 39 MPa

Temperature:	400°C		
Residence Time (sec)	45	90	180
% CH_3NO_2 Recovered	57	43	16

Temperature:	500°C		
Residence Time (sec)	15	30	60
% CH_3NO2 Reserved	8	<1	<1
% NO_3^-, NO_2^-			<0.005

Temperature:	580°C		
Residence Time (sec)	9.4	19	38
% CH_3NO_2 Recovered	<1	<1	<1
% NO_3^-, NO_2^-			<0.2

287

can be represented by a free-radical mechanism appropriately modified to include the effects of pressure and the high concentration of water. This hypothesis forms a powerful basis for extrapolation of kinetic data to other compounds, with significantly less experimental work needed than would be required for determinations of the kinetics of destruction of every compound to be addressed with this technology. In addition, the particular data obtained and the mechanisms developed in this study are part of this basis for extrapolation to other compounds.

The current mechanism is derived from a gas-phase mechanism by additions and modifications that are well-supported and are of the type expected for the conditions of supercritical water oxidation. Further modifications are needed and will be made to the mechanism as our understanding of molecular interactions in supercritical water increases. The carbon monoxide results show a significant and, as yet, undescribed effect of the non-ideal nature of supercritical water. As the temperature is increased and the system becomes closer to ideal, the mechanism developed here shows good agreement with experimental results. Experimental results on ammonium perchlorate and nitromethane suggest that supercritical water processes may be effective for destruction of at least some propellants and explosives.

ACKNOWLEDGMENTS

We are grateful to the Department of Energy Office of Technology Development and to the Air Force Engineering and Services Center for their support of programs in supercritical water methods of waste destruction.

The project team is: J. H. Atencio, G. R. Brewer, S. J. Buelow, R. B. Dyer, E. O. Ferdinand, B. R. Foy, D. L. Hanson, D. M. Harradine, J. L. Lyman, K. A. Mays, R. D. McFarland, W. J. Parkinson, and C. Wohlberg. G. E. Streit developed the detailed chemical kinetics model. Experimental data on CO and CH_4 were obtained at MIT by J. W. Tester and his graduate students.

REFERENCES

1. L. Haar, J. S. Gallagher, and G. S. Kell, NBS/NRC Steam Tables, Hemisphere Publishing Company, Washington, 1984.

2. For example, T. W. de Loos, J. H. van Dorp, and R. N. Lichtenthaler, "Phase equilibria and critical phenomena in fluid (n-alkane + water) at high pressures and temperatures," Fluid Phase Equil. 1983, 10, 279-287.

3. T. W. Leland, J. J. McKetta, Jr., and K. A. Kobe, "Phase equilibrium in 1-butene-water system and correlation of hydrocarbon-water solubility data," Ind. Eng. Chem. 1955, 47, 1265-1271.

4. For example, Z. Alwani and G. Schneider, "Druckeinfluss auf die Entmischung fluessiger Systeme. VI. Phasengleichgewichte und kritische Erscheinungen im System Benzol-H2O zwischen 250 und 368°C bis 3700 bar," Ber. Bunsenges. Phys. Chem. 1967, 71, 633-638.

5. R. Jockers and G. M. Schneider, "Fluid mixtures at high pressures. Fluid phase equilibria in the systems fluorobenzene + water, 1,4-difluorobenzene + water, and 1,2,3,4-tetrahydronaphthalene + decahydronaphthalene(trans) + water up to 360 MPa," Ber. Bunsenges. Phys. Chem. 1978, 82, 576-582.

6. M. L. Japas and E. U. Franck, "High pressure phase equilibria and PVT data of the water-oxygen system including water-air to 673 K and 250 MPa," Ber. Bunsenges. Phys. Chem. 1985, 89, 1268-1275.

7. M. L. Japas and E. U. Franck, "High pressure phase equilibria and PVT data of the water-nitrogen system to 673 K and 250 MPa," Ber. Bunsenges. Phys. Chem. 1985, 89, 703-800.

8. T. M. Seward and E. U. Franck, "The system hydrogen-water up to 440°C and 2500 bar pressure," Ber. Bunsenges. Phys. Chem. 1981, 85, 2-7.

9. For example, (a) R. Koningsveld, L. A. Kleintjens, and G. A. M. Diepen, "Solubility of solids in supercritical solvents. I. General Principles," Ber. Bunsenges. Phys. Chem. 1984, 88, 848-855. (b) J. C. Tanger IV and K. S. Pitzer, "Thermodynamics of NaCl-H$_2$O: A new equation of state for the near-critical region and comparisons with other equations for the adjoining regions," Geochim. Cosmochim. Acta 1989, 53, 973-987.

10. B. Subramaniam and M. A. McHugh, "Reactions in supercritical fluids--A review," Ind. Eng. Chem. Process Des. Dev. 1986, 25, 1-12.

11. (a) T. B. Thomason and M. Modell, "Supercritical water destruction of aqueous wastes," Hazardous Waste 1984, 1, 453-467. (b) K. C. Swallow, W. R. Killilea, K. C. Malinowski, and C. N. Stazak, "The Modar Process for the Destruction of Hazardous Organic Wastes--Field Test of a Pilot-Scale Unit," Waste Management 1989, 9, 19. (c) J. B. Johnston, R. E. Hannah, V. L. Cunningham, B. P. Daggy, F. J. Sturm, and R. M. Kelly, "Destruction of Pharmaceutical and Biopharmaceutical Wastes by the Modar Supercritical Water Oxidation Process," Bio/Technology 1988, 6, 1423.

12. (a) S. H. Timberlake, G. R. Hong, M. Simson, and M. Modell, "Supercritical water oxidation for wastewater treatment: Preliminary study of urea destruction, SAE Technical Paper 820872, Twelfth Intersociety Conference on Environmental Systems, San Diego, California, July 1982. (b) R. K. Helling and J. W. Tester, "Oxidation kinetics of carbon monoxide in supercritical water," J. Energy Fuels

1987, 1, 417-423. (c) R. K. Helling and J. W. Tester, "Oxidation kinetics of simple compounds and mixtures in supercritical water: Carbon monoxide, ammonia, and ethanol," Env. Sci. Tech. 1988, 22, 1319-1324. (d) H. H. Yang and C. A. Eckert, "Homogeneous catalysis in the oxidation of p-chlorophenol in supercritical water," Ind. Eng. Chem. Research 1988, 27, 2009-2014.

13. (a) M. A. Abraham and M. T. Klein, "Pyrolysis of benzylphenylamine neat and with tetralin, methanol, and water solvents," Ind. Eng. Chem. Prod. Res. Dev. 1985, 24, 300-306. (b) M. A. Abraham and M. T. Klein, "Solvent effects during the reaction of coal model compounds," ACS Symp. Ser. 1987, 329 (Supercrit. Fluids: Chem. Eng. Princ. Appl.), 67-76. (c) D. S. Ross, G. P. Hum, T. C. Miin, T. K. Green, and R. Mansani, "Isotope effects in supercritical water: Kinetic studies of coal liquefaction," ACS Symp. Ser. 1987, 329 (Supercrit. Fluids: Chem. Eng. Princ. Appl.), 242-250. (d) G. V. Deshpande, G. D. Holder, and Y. T. Shah, "Effect of solvent density on coal liquefaction kinetics," ACS Symp. Ser. 1987, 329 (Supercrit. Fluids: Chem. Eng. Princ. Appl.), 251-265. (e) T. J. Houser, D. M. Tiffany, Z. Li, M. E. McCarville, and M. E. Houghton, "Reactivity of some organic compounds with supercritical water," Fuel 1986, 329, 827-832. (f) T. J. Houser, C.-C. Tsao, J. E. Dyla, M. K. Van Atten, and M. E. McCarville, "The reactivity of tetrahydroquinoline, benzylamine and bibenzyl with supercritical water," Fuel 1989, 68, 323-327.

14. For example, M. J. Antal, Jr., A. Brittain, C. DeAlmeida, W. Mok, and S. Ramayya, "Catalyzed and uncatalyzed conversion of cellulose biopolymer model compounds to chemical feedstocks in supercritical solvents," Energy Biomass Wastes 1987, 10th, 865-877.

15. For example, G. D. Rappe and W. L. Schwoyer, "Evaluating an Enclosed System for Containment and Destruction of Hazardous and Toxic Wastes," Presented at the American Institute of Chemical Engineers Summer Meeting, Boston, Massachusetts, August 1986.

16. C. K. Rofer and G. E. Streit, "Phase II Final Report: Oxidation of Hydrocarbons and Oxygenates in Supercritical Water," LA-11700-MS (DOE/HWP-90), Los Alamos National Laboratory, Los Alamos, NM (1989).

17. S. J. Buelow, R. B. Dyer, C. K. Rofer, J. H. Atencio, and J. D. Wander, "Advanced techniques for soil remediation: Destruction of propellant components in supercritical water," Proceedings of the JANNAF Workshop on Alternatives to Open Burning/Open Detonation of Propellants and Explosives, in press.

18. M. Modell, "Supercritical Water Oxidation," in Standard Handbook of Hazardous Waste Treatment and Disposal, edited by H. M. Freeman, McGraw-Hill Book Company, New York, 1989, pp. 8.153-8.168.

19. (a) A. R. Wilhelmi and P. V. Knopp, "Wet air oxidation--An alternative to incineration," Chem. Eng. Prog. 1979 (August), xx, 46-52. (b) T. L. Randall and P. V. Knopp, "Detoxification of specific organic substances by wet oxidation," J. Water Pollution Control 1980, 52, 2117-2130. (c) M. K. Conditt and R. E. Sievers, Microanalysis of reaction products in sealed tube wet air oxidations by capillary gas chromatography," Anal. Chem. 1984, 56, 2620-2622. (d) M. J. Dietrich, T. L. Randall, and P. J. Cannery, "Wet air oxidation of hazardous organics in wastewater," Env. Prog. 1985, 4, 171-177.

20. C. F. Melius, N. Bergan, and J. E. Shepherd, "Effects of water on combustion kinetics at high pressure," 23rd International Symposium on Combustion, July 22-27, 1990, in press.

21. For example, (a) D. E. Hoare, G. B. Peacock, and G. R. D. Ruxton, "Efficiency of surfaces in destroying hydrogen peroxide and hydroperoxyl radicals," Trans. Farady Soc. 1967, 63, 2498-2503. (b) S. Mahajan, W. R. Menzies, and L. F. Albright, "Partial oxidation of light hydrocarbons. 1. Major differences noted in various tubular reactors," Ind. Eng. Chem., Process Des. Dev. 1977, 16, 271-274. (c) I. A. Vardanyan and A. B. Nalbandyan, "On the mechanism of thermal oxidation of methane," Int. J. Chem. Kinet. 1985, 17, 901-923.

T. J. HOUSER
C. C. TSAO

Reactions of Organic Compounds with Supercritical Water Involving Chemical Oxidation

ABSTRACT

Initial interest in the reactions of supercritical water (SW) with organic compounds developed because of the possible use of supercritical fluid extraction of coal to obtain cleaner, more versatile fluid products. At or above critical conditions water can react with many organic functional groups, which for some compounds leads to products significantly different than those obtained from pyrolysis. Thus, water may provide a means of removing heteroatoms as well as to function as a solvent. This paper describes the results of reactions of benzaldehyde, benzyl alcohol and benzylidenebenzylamine ($C_6H_5CH_2N=CHC_6H_5$) with SW since these compounds were determined to be major intermediates in the reaction of benzylamine with SW. In addition, the influences of ammonia and dihydroanthracene on the product distributions and extents of reaction were determined. Of particular interest is the reaction pathway leading to the formation of benzene and related products which requires an oxidation/hydrolysis sequence, terminated by decarboxylation to form hydrocarbons. The mechanistic significance of the benzene/toluene product yield ratios will be discussed.

INTRODUCTION

The possible use of supercritical fluid extraction (SFE) of coal to obtain cleaner, more versatile fluid products has been of significant interest. Some fluids have the opportunity to participate as reactants at process conditions, which may yield extracts of very different compositions than those obtained from other treatments and which will be dependent on the fluid used. Thermodynamic consideration of SFE leads to the prediction that the enhanced solubility (volatility) of the solute may be several orders of magnitude [1-3]. Thus, this method combines many of the advantages of distillation with those of extraction. However, despite the interest in SFE, only a few studies have reported the basic chemistry that may be taking place during coal extraction at these conditions [4-8].

The results which are being reported are concerned with nitrogen removal from organic model compounds thought to be representative of structures found in fossil fuels. Because of the difficulty of removing heterocyclic nitrogen, experiments were initiated by extensively examining the reactivities of quinoline and isoquinoline, as well as brief examinations of the reactivities of other compounds [4]. The selection of water as the fluid was based on its physical and chemical properties [9] and on the

T. J. Houser and C. C. Tsao, Chemistry Department, Western Michigan University, Kalamazoo, Michigan, U.S.A.

observation that a few studies of SFE of coal using water as the fluid have given encouraging results [10,11]. Zinc chloride was chosen as a catalyst because of its reported catalytic activity for hydrocracking aromatic structures [12]. This paper discusses the results of a study of the reactions of supercritical water (SW) with organic compounds that were found, or postulated, to be intermediates in the reaction of isoquinoline with SW. Since the isoquinoline reaction produced significant yields of ethyl benzene and o-xylene [4] it was assumed that benzylamine (BA) would be representative of the intermediate structure formed after the initial bond rupture in the heterocycle. It was found that benzaldehyde, benzylidenebenzylamine (BBA) and benzyl alcohol were intermediates in the BA-SW reaction [5], the yields of which decreased as the benzene yield increased. Thus these are the primary subjects of the current study with additional information obtained from the chemical behavior of bibenzyl and benzoic acid.

EXPERIMENTAL

The experiments were carried out in a small (47 cm^3) stainless steel, batch reactor, which was not equipped for the collection of gaseous products for analysis. The reactor was loaded with about 2.00 g of an organic compound. Water (10 ml) was added for the SW experiments to produce the desired pressure at reaction temperature, catalysts were added as needed, then the reactor was purged with argon and bolted closed using a copper gasket. The reactor was placed in a fluidized sand bath furnace for the required reaction time, about 15 minutes was required to reach 375C. Following reaction, the vessel was air cooled, opened, the reaction mixture removed and the water and organic layers separated. Portions of methylene chloride solvent were used to rinse the reactor and extract the water layer. These portions were combined with the organic layer and additional solvent added to a standard volume for quantitative determinations made gas chromatographically using peak area calibrations from known solutions. The components for these solutions were identified mass spectrometrically.

There were certain limitations on the g.c.-m.s. determinations: Some components could not be separated completely and these are reported as a total yield of mixture using an average calibration factor. Some products are reported as an isomer of a probable structure as deduced from the molecular weight and m.s. fragmentation pattern. Finally, many of the higher molecular weight minor products could be measured only with a low degree of precision by g.c. and calibration factors were estimated.

RESULTS AND DISCUSSION

The previous study of the isoquinoline - SW reaction [4] indicated that following the rupture of the CN bond in the 1-2 or 2-3 position, the nitrogen portion would undergo hydrolysis and decarboxylation (similar to that observed for benzonitrile to yield benzene) while the carbon end was either capped directly, or shortened and capped, by hydrogen thus producing toluene, ethylbenzene or o-xylene, the major volatile products. To help substantiate this proposed sequence of reactions, the benzylamine -SW reaction was studied [5]. At 450C and longer reaction times the significant products were as shown by Equation 1:

$$PhCH_2NH_2 \xrightarrow{SCW} PhCH_3 + PhH + Ph-Ph + Ph-PhCH_3 \qquad (1)$$

where (Ph = $C_6H_5\cdot$). However, at milder conditions benzaldehyde, benzyl alcohol and benzylidenebenzylamine ($C_6H_5CH_2N=CHC_6H_5$) were observed as intermediates. Thus, it was necessary to determine how they contributed to the formation of the final products. The data reported in Tables I-IV are representative of the results obtained. Ammonia was added in some experiments since it would be present in the denitrogenation of amines.

BENZYL ALCOHOL

TABLE I shows that SW by itself had relatively little influence on the on the pyrolysis reaction, both giving low yields of volatile products and

TABLE I - BENZYL ALCOHOL - SW REACTION[a]

Water Pressure (psi)	0	3870	3870	3870
Added NH_3 (Molar)	0	0	2	6
Volatile Products (%Yield)				
Benzene	2.8	2.6	19	21
Toluene	24	20	42	44
Benzaldehyde	0	6.0	9.5	5.5
Biphenyl	0.1	0	2.3	2.6
168 Isomers[b]	3	3	4	5
182 Isomers[b]	8	11	2	2
DHP/Silbene[b]	0	0	1.6	1.4
BBA	-	-	1.2	1.6
Above 190[b]	8	-	1	3

a. All experiments were at 400C for 3 hrs, all extents of reaction were about 100% and the two experiments without ammonia gave black tarry products, the others gave clear yellow solutions.

b. The numbers represent molecular weights of mixtures, DHP is dihydrophenanthrene.

significant amounts of black char/tar. The added ammonia suppressed the char/tar formation, producing clear yellow solutions and significantly higher yields of volatile products. It is believed that the most important observations are: (a) that for all experiments toluene is formed in the highest yield and (b) the presence of SW does promote the formation of benzaldehyde which, as will be shown later, would lead to higher benzene yields. The second observation is consistent with other data that have shown SW capable of removing hydrogen from tetralin [4] and benzylamine (to be discussed later).

BENZYLIDENEBENZYLAMINE (BBA)

The formation of BBA in the BA - SW reaction was probably from the following series of reactions:

$$PhCh_2NH_2 \xrightarrow[-H_2]{SCW} PhCH=NH \xrightarrow{SCW} PhCHO + NH_3 \qquad (2)$$

$$PhCHO + PhCH_2NH_2 \rightarrow PhCH=NCH_2Ph + H_2O \qquad (3)$$

since benzaldehyde was also found as an intermediate in this reaction. The data in TABLE II show that BBA - SW reaction produces benzene and benzene related products in somewhat larger yields than toluene. However, since some benzaldehyde is also produced the above reaction must be somewhat reversible. It appears that benzene and other products formed from phenyl radicals have benzaldehyde as their precursor, the evidence for which will be given later. An important point to note is that the addition of a hydrogen donor, dihydroanthracene (DHA), promotes the formation of toluene at the expense of benzene and other species which form from benzaldehyde. The following reaction sequence is a possible explanation:

$$PhCH=NCH_2Ph \xrightarrow{DHA} (PhCH_2)_2NH \xrightarrow{DHA} 2\ PhCH_3 + NH_3 \qquad (4)$$

TABLE II - BENZYLIDENEBENZYLAMINE - SW REACTION[a]

Time (h)	1	3	6	9	3
Additive	0	0	0	0	2gDHA
Volatile Products (%Yield)					
Benzene	22	29	35	34	11
Toluene	24	26	28	29	54
Benzaldehyde	22	11.0	3.4	2.0	0
Benzyl Alcohol	2.8	2.4	0.2	0.1	0
Benzamide	0.6	0.4	0	0	0
Biphenyl	4.1	6.4	7.3	7.2	0.4
168 Isomers[b]	4	6	7	7	-
Fluorene	0.1	0.4	0.6	0.8	-
Bibenzyl	2.7	2.6	2.0	1.9	-
Benzophenone	1.3	1.9	1.0	0.8	-
182-196 Mixture[b]	4	4	2	2	-
Above 200[b]	6	6	8	8	-

a. All experiments were at 400C and about 3870 psi water pressure. The extents of reaction were all above 95%. All solutions were clear.

b. The numbers represent molecular weights of product mixtures.

Thus, it appears that rupture of the single CN bond leads to toluene while the C=N segment can be hydrolysed by the reverse of the imine formation reaction, possibly by the following reactions:

$$PhCH=N-CH_2Ph + H_2O \rightarrow \left[\underset{\underset{OH}{|}}{Ph-\overset{\overset{H}{|}}{C}}-\overset{\overset{H}{|}}{N}-CH_2Ph \right] (A) \tag{5}$$

$$A \begin{cases} \rightarrow PhCONH_2 + PhCH_3 & (6) \\ \rightarrow PhCHO + PhCH_2NH_2 & (7) \end{cases}$$

$$PhCONH_2 + H_2O \rightarrow PhCO_2H + NH_3 \tag{8}$$

$$PhCHO + H_2O \rightarrow \left[Ph-\underset{\underset{OH}{|}}{\overset{\overset{H}{|}}{C}}-OH \right] (B) \rightarrow PhCO_2H + H_2 \tag{9}$$

The BA can react as in Equation 3 or form benzyl and amine radicals from a CN rupture which then can abstract H atoms to form toluene and NH_3. Benzoic acid can decarboxylate rapidly in the presence of NH_3 to form benzene as shown by the data in TABLE III.

TABLE III - BENZOIC ACID - SW REACTION*

Time (h)	3	4	4
Water Pressure (psi)	0	3870	3870
Added NH$_3$	0	0	2 M
% Reaction	95	58	100
Volatile Products (%Yield)			
Benzene	80	79	92
Phenol	1.5	0.4	0.1
Biphenyl	3.1	0.4	0.2
* All experiments were at 400C.			

BENZOIC ACID

It has been assumed that benzene is formed from the decarboxylation of benzoic acid although it had not been observed as an intermediate in the reactions studied earlier. The data in TABLE III show that benzoic acid does react about as expected. However, while SW appears to slow the reaction compared to pyrolysis, ammonia acts as a catalyst to produce an almost quantitative yield of benzene.

BENZALDEHYDE

Again it would appear that SW has relatively little effect on the pyrolysis of benzaldehyde. The data in TABLE IV show that the extents of reaction and product distributions are very similar (except the benzyl alcohol yields are increased somewhat with SW) for the reactions with and without water. However, with the addition of ammonia the reaction rates are increased as well as the toluene yields. The most important observation is that the major product is benzene indicating an oxidation/decarboxylation sequence is predominant. Clearly the presence of a source of hydrogen (DHA) promotes the benzyl alcohol sequence to form toluene at the expense of benzene. To a lesser degree ammonia has a similar effect but the mechanism by which this occurs would be very speculative.

The conclusion that can be drawn from the product distribution is that benzene and other products formed from phenyl radicals are produced through the oxidation of benzaldehyde, Equation 9. However, in that process since hydrogen is produced it is possible to reduce some unreacted benzaldehyde to the alcohol from which toluene and products formed from benzyl radicals are obtained, with some reversibility in the oxidation/reduction part of the sequence.

Another possibility is that the Cannizzaro disproportionation, through intermediate B is taking place, Equation 10. However, this would be

$$B + PhCHO \rightarrow PhCO_2H + PhCH_2OH \tag{10}$$

expected to lead to equal molar yields of benzoic acid and benzyl alcohol or their respective subsequent products, unless followed by hemiacetal formation and disproportionation, Equation 11:

$$PhCH_2OH + PhCHO \rightarrow \left[PhCH_2O\text{-}\overset{\text{OH}}{\underset{\text{H}}{C}}\text{-}Ph \right] \rightarrow PhCH_3 + PhCO_2H \tag{11}$$

TABLE IV – BENZALDEHYDE – SW REACTION[a]

Time (h)	1[b]	6[b]	1	3	6	1	3	6	1	3	9
NH_3 (M)	0	0	0	0	0	2M	2M	2M	6M	2M	2M
DHA (g)	0	0	0	0	0	0	0	0	0	2	2
% Reaction	29	76	25	53	71	73	85	93	94	89	99
Volatile Products[c] (%Yield)											
Benzene	70	54	36	49	50	32	41	44	28	16	18
Toluene	6.7	7.3	5.9	8.3	6.4	11	17	21	12	39	38
Benzyl Alcohol	1.7	0	6.8	1.2	0.9	2.1	2.8	2.5	3.4	–	–
Benzoic Acid	10	6.9	14	7.5	4.7	0	0	0	0	–	–
Benzamide	–	–	–	–	–	1.2	1.0	0.7	2.6	–	–
Biphenyl	11	14	3.4	6.1	8.3	4.8	9.2	12	3.4	–	–
168 Isomers	6	9	2	3	4	2	4	6	3	–	–
166,180–196 Mixture	19	8	10	8	5	6	7	5	5	–	–
Benzylidenebenzylamine	–	–	–	–	–	2.0	4.3	4.0	6.1	–	–
Above 200	1	3	<1	2	4	20	7	5	13	–	–

a. All experiments were at 400C.

b. These experiments were pyrolyses, the others had about 3870 psi water pressure.

c. The numbers represent molecular weights.

The stoichiometry of the combined reactions 9-11 results in benzene vs. toluene (and their related products) formed in a two to one ratio.

MECHANISM CONSIDERATIONS

The formation of biphenyl and 168 isomers (methyl biphenyls) as well as benzene demonstrate that phenyl radicals exist as intermediates. The formation of bibenzyl and other 182 isomers (benzyl toluenes) and the 168 isomers as well as toluene demonstrate the presence of benzyl radicals. The previous results from experiments with bibenzyl and SW [5], including some with added NH_3 [13], show conclusively that once benzyl radicals form, they abstract hydrogen or combine with other radicals resulting in very little (if any) oxidation/decarboxylation to benzene. These results from bibenzyl in addition to those obtained from BA, BBA and BBA plus DHA lead to the following conclusions: (a) The CN bond in the reactant is necessary for the hydrolysis/oxidation sequence to be initiated. (b) If a CN single bond exists in an aromatic side chain it may rupture to form a radical that is capped by hydrogen producing an alkylated aromatic, or hydrogen can be removed to form a CN multiple bond. (c) Once formed, the CN multiple bond can undergo the oxidation/decarboxylation process. (d) These reactions are somewhat reversible. (e) Ammonia catalyzes the hydrolysis/oxidation and decarboxylation processes, probably through amide formation. (f) Finally, it must be concluded that SW can facilitate the removal or transfer of hydrogen from many reactants and intermediates as evidenced by the results of tetralin in SW [4], benzaldehyde sequence of reactions from the BA-SW reaction [5], benzamide from benzaldehyde-SW with added ammonia (TABLE IV) and for many of the hydrolytic/oxidation sequences leading to the observed products to be possible.

REFERENCES

1. N. Gangoli and G. Thodos, Ind. Eng. Chem., Prod. Res. Dev., 16, 208, (1977).
2. D.F. Williams, Chem. Eng. Science, 36, 1769, (1981).
3. J.C. Whitehead and D.F. Williams, J. Inst. Fuel, 182, (1975).
4. T.J. Houser, D.M. Tiffany, Z. Li, M.E. McCarville and M.E. Houghton, Fuel, 65, 827, (1986).
5. T.J. Houser, C.C. Tsao, J.E. Dyla, M.K. VanAtten and M.E. McCarville, Fuel, 68, 323, (1989).
6. M.A. Abraham and M.T. Klein, Ind. Eng. Chem. Prod. Res. Dev., 24, 300, (1985).
7. S.H. Townsend and M.T. Klein, Fuel, 64, 635, (1985).
8. J.R. Lawson and M.T. Klein, Ind. Eng. Chem. Fund., 24, 203, (1985).
9. E.U. Frank, Endeavour, 27, 55, (1968).
10. G.V. Deshpande, G.D. Holder, A.A. Bishop, J. Gopal and I. Wender, Fuel, 63, 956, (1984).
11. V.I. Stenberg, R.D. Hei, P.G. Sweeny and J. Nowak, Preprints, Div. Fuel Chem., Am. Chem. Soc., 29(5), 63, (1984).
12. S.S. Salim and A.T. Bell, Fuel, 63, 469, (1984).
13. C.C. Tsao and T.J. Houser, unpublished results.

WILLIAM M. COPA
JOSEPH A. MOMONT
DAVID A. BEULA

The Application of Wet Air Oxidation to the Treatment of Spent Caustic Liquor

ABSTRACT

Wet air oxidation has been applied to the treatment of spent caustic liquors which are generated in the petrochemical and refinery industries. Ethylene is manufactured in the petrochemical industry using a thermal cracking process. The feed stock to the cracker is ethane, propane, butane or naphtha. The cracked gas is contacted in a scrubber with an aqueous sodium hydroxide solution for the removal of acid gases. The most abundant acid gases are carbon dioxide and hydrogen sulfide which are retained in the caustic scrubbing liquor as sodium carbonate and sodium sulfide. The caustic scrubbing liquor also removes mercaptans, phenols, and other organic compounds which are soluble or emulsified in the caustic liquor. A purge stream of spent caustic scrubbing liquor is discharged from the scrubbing tower. This spent caustic scrubbing liquor is classified as a D003 (Reactive Sulfide) Resource Conservation and Recovery Act (RCRA) hazardous waste. As such, these spent caustic liquors must be treated to a standard of "deactivation to remove the characteristic of reactivity". Deactivation of reactive sulfide waste consists of chemically converting the sulfide to relatively inert sulfur, to insoluble metallic sulfide salts, or to soluble sulfates.

In the refinery industry, caustic is used to extract acidic components from liquid products. This spent caustic also contains sulfides, mercaptans, and phenols as well as emulsified hydrocarbons.

William M. Copa, Vice President—Technical Services, Zimpro Passavant, Rothschild, Wisconsin, U.S.A.

Joseph A. Momont, Materials Specialist, Zimpro Passavant, Rothschild, Wisconsin, U.S.A.

David A. Beula, Senior Process Engineer, Zimpro Passavant, Rothschild, Wisconsin, U.S.A.

Wet air oxidation can be used to accomplish deactivation of the spent caustic liquors. In the wet air oxidation process, the sulfur from the inorganic sulfides and mercaptans is oxidized to sulfate, thereby removing the reactive characteristic.

Wet air oxidation refers to a process which involves an aqueous phase oxidation of organic and inorganic materials at elevated temperatures and pressures. In wet air oxidation, oxidation and hydrolysis reactions occur at temperatures in the range of 150° to $320^{\circ}C$ (302° to $608^{\circ}F$) and at corresponding pressures of 2068 to 20,684 kPa (300 to 3000 psig). Oxygen, from compressed air or pressurized oxygen gas, serves as the oxidizing agent.

The treatment of spent caustic liquors by the wet air oxidation process is discussed in the present paper. Data are presented on the wet air oxidation of actual spent caustic liquors. The conversion of sulfides to sulfates in the wet air oxidation process is documented.

INTRODUCTION

Spent caustic wastewaters that are generated in the petrochemical and refinery industries are classified as Resource Conservation Recovery Act (RCRA) hazardous wastes[1]. This classification stems from the sulfide content of these wastewaters which is typically at a concentration of several thousand milligrams per liter. These wastes are classified as characteristically hazardous and exhibit a reactivity characteristic (sulfide bearing waste). The RCRA hazardous waste code for these spent caustic wastewaters is D003.

The United States Environmental Protection Agency has promulgated a treatment standard of "Deactivation to Remove the Characteristic of Reactivity". Deactivation of reactive sulfides can be achieved by chemical conversion of the sulfide to elemental sulfur, to insoluble metallic sulfides, or by oxidation to soluble sulfates.

Wet air oxidation is a process that can be used to achieve the oxidation of reactive sulfide to soluble sulfate. This conversion is achieved at temperatures of 150 to $200^{\circ}C$ and corresponding pressures of 150 to 400 psig. In the wet air oxidation process, the sulfides and mercaptans that are present in the spent caustic liquors are quantitatively oxidized to sulfates. In addition, a substantial portion of the organic constituents, e.g. phenols, oils, and polymers, are also removed by oxidation reactions.

The present paper discusses the chemical conversions that take place in the wet air oxidation of spent caustics. This paper also reports on the characteristics of various spent caustic wastewater and the treatment standards that can be achieved by the application of the wet air oxidation technology.

PROCESS DESCRIPTION

The wet air oxidation process is an aqueous phase oxidation of organic and inorganic constituents. The oxidation reactions are affected at elevated temperatures and pressures, using oxygen from air as the oxidizing agent. The temperature range that has been applied in commercial wet air oxidation systems is approximately 300 to 608°F (149 to 320°C) with corresponding pressures of 300 to 3000 psig (2068 to 20684 kPa).

The basic wet air oxidation process flow diagram is shown in Figure 1. In this process, a wastewater which contains the oxidizable constituents is brought up to system pressure using a high pressure pump. Compressed air or oxygen gas is introduced into the pressurized wastewater stream. The mixture of gas and wastewater is heated in a process heat exchanger by heat exchange with the oxidized effluent. An external source of heat is used to initiate the wet air oxidation process or to sustain the oxidation temperature if insufficient heat of reaction is released in the wet air oxidation reaction.

After heating, the mixture of gas and wastewater flows into the reactor where it is detained for a period of time which is sufficient to complete the desired degree of oxidation. The reactor is a vertical bubble column pressure vessel which is sized to provide the desired hydraulic residence time. The wet air oxidation reactions are exothermic and raise the temperature of the mixture to the desired operating temperature (i.e. 300 to 608°F). The hot oxidized effluent that flows from the reactor is directed into the process heat exchanger to preheat the incoming mixture and cool the oxidized effluent. An optional water cooler may be used to further cool the oxidized effluent. After cooling, the oxidized effluent passes through a pressure control valve (PCV) and is directed into a separator where the non-condensible gases separate from the oxidized liquid phase.

The wet air oxidation process has been applied to the treatment of a wide variety of wastewaters[2].

CHARACTERISTICS OF SPENT CAUSTICS

An ethylene cracker produces ethylene and other olefins from hydrocarbons including ethane, propane, and butane and also from liquid naphtha. The produced olefin gas is usually contacted with an amine scrubbing liquid for ultimate recovery of sulfur. After amine scrubbing, the produced gas is scrubbed in one or more stages with aqueous sodium hydroxide to accomplish the removal of acid components which include carbon dioxide, hydrogen sulfide,

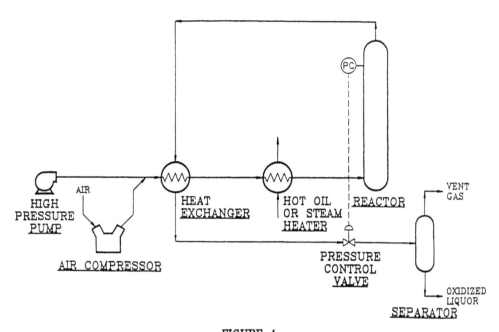

FIGURE 1

WET AIR OXIDATION FLOW DIAGRAM

mercaptans, and phenols. The produced gas is finally
scrubbed with water to eliminate any sodium hydroxide
carryover. The caustic scrubbing liquid that is purged
from the sodium hydroxide scrubbing stages is usually
combined with the final washwater[3] and constitutes the
spent caustic stream from an ethylene cracker. The spent

caustic from the production of ethylene contains variable amounts of sodium hydroxide, sodium carbonate, mercaptans, phenols, and other miscellaneous organics. The chemical analyses of five spent caustic liquors from different ethylene units are shown in TABLE I. These analyses of spent caustic liquors show that the excess sodium hydroxide is a variable and is typically determined by the operation of the sodium hydroxide scrubber.

In the refinery industry, many refined liquid products are extracted with a sodium hydroxide solution for removal of acidic components, e.g. hydrogen sulfide, phenols, and cresols. The sodium hydroxide extract contains the extracted acidic components and is saturated with the hydrocarbon stream that was initially extracted. These extracts constitute a spent caustic from the refinery industry. The analyses of several of these refinery spent caustics is shown in TABLE II. These analyses show that the refinery spent caustics have a substantially higher organic content and Chemical Oxygen Demand (COD) than the ethylene spent caustics.

WET AIR OXIDATION OF SPENT CAUSTIC

In the wet air oxidation of spent caustics, sulfides, mercaptans, and a portion of the organic constituents are oxidized. The oxidation of the inorganic sulfides proceeds according to the following reactions:

$$Na_2S + 2O_2 \rightarrow Na_2SO_4 \tag{1}$$

$$NaHS + 2O_2 + NaOH \rightarrow Na_2SO_4 + H_2O \tag{2}$$

The oxidation of inorganic sulfide to sulfate is mainly temperature dependent. The mechanism of this reaction may involve an intermediate formation of thiosulfate. The thiosulfate is subsequently oxidized to sulfate. Evidence for this type of mechanism is shown in TABLE III and Figure 2. These data show that the sulfide concentration is rapidly diminished even at a temperature of 100°C. The increasing concentration of sulfate, as a function of temperature and time, is shown in Figure 2. The conversion of sulfide to sulfate is not complete at the lower temperatures, whereas, the destruction of sulfide, i.e. the oxidation of sulfide to perhaps thiosulfate, is completed at a temperature of 100°C or less.

Mercaptans are oxidized according to the following reaction:

TABLE 1
CHARACTERISTIC OF SPENT CAUSTIC
FROM ETHYLENE PRODUCTION

COMPONENT	SPENT CAUSTIC SAMPLES				
	#1	#2	#3	#4	#5
COD, g/l	41.7	45.5	28.4	26.2	33.7
TOTAL SOLIDS, g/l	110	216	108	82.1	--
ASH, g/l	109	207	107	57.5	--
SOLUBLE CHLORIDE, g/l	88.9	68.1	42.4	13.5	--
pH	13.78	14.03	13.86	13.48	13.54
TOTAL SULFUR, g/l	26.5	17.8	13.5	--	--
Na_2S, g/l	9.4	4.4	3.1	0.15	1.76
NaHS, g/l	36.4	28.8	20.2	2.99	30.6
NaSR, g/l	--	--	--	--	1.43
NaOH, g/l	42.7	54.6	60.4	36.9	15.6
Na_2CO_3, g/l	13.8	14.2	6.4	13.6	12.1

TABLE II
CHARACTERISTICS OF
PETROLEUM REFINERY SPENT CAUSTIC

	Spent Caustic Samples	
	#1	#2
BOD, mg/l	74,600	-
COD, mg/l	151,000	108,100
Total Phenols, mg/l	20,200	15,510
Total Sulfur, mg/l	12,260	3,580
Sulfate Sulfur, mg/l	1,890	570
Sulfide Sulfur, mg/l	1,440	
3,010(1)		
pH	12.6	13.0
Soluble Chloride, mg/l	8,280	1,510
Total Solids, g/l	232.3	88.6
Total Ash, g/l	201.8	57.1

(1) Determined by difference (Total Sulfur - Sulfate Sulfur)

TABLE III

THE WET AIR OXIDATION OF
SULFIDE AT A TEMPERATURE OF $100^{\circ}C$

Time at Temperature, Min.	Sulfide as S, mg/l
Influent	1260
15	428
30	12
45	<1
60	<1

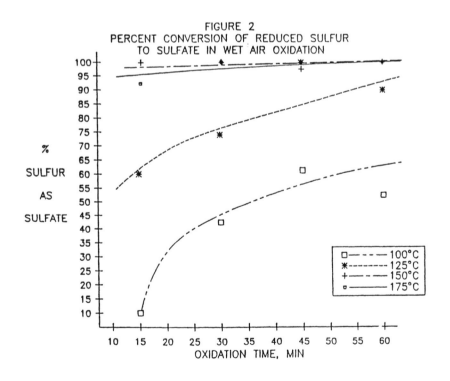

FIGURE 2
PERCENT CONVERSION OF REDUCED SULFUR
TO SULFATE IN WET AIR OXIDATION

$$NaSR + 3O_2 + 2NaOH \rightarrow Na_2SO_4 + NaO\overset{O}{\overset{\|}{C}} - R' + 2H_2O \quad (3)$$

where the organically bound sulfur is converted to sulfate and the organic portion of the mercaptan, i.e. R, is converted to the corresponding carboxylic acid anion. If the mercaptan in the spent caustic is ethyl mercaptan, the reaction is

$$NaSC_2H_5 + 3O_2 + 2NaOH \rightarrow Na_2SO_4 + NaO\overset{O}{\overset{\|}{C}} - CH_3 + 2H_2O \quad (4)$$

Phenols are usually present in the spent caustic from ethylene production and are a major constituent in refinery spent caustic. The destruction of phenol, i.e. C_6H_5OH, occurs at rather low temperatures and is essentially complete at $150^\circ C$ (see TABLE IV). The wet air oxidation destruction of total phenols at various temperatures is shown graphically in Figure 3. It is anticipated that the higher molecular weight phenols react more slowly but that a high destruction of substituted phenols can be accomplished at temperatures of $175^\circ C$ and above.

TABLE IV

THE WET AIR OXIDATION OF PHENOL
(i.e. C_6H_5OH) AS A FUNCTION OF TEMPERATURE

Oxidation Temperature[1]	Phenol, mg/l[2]
Influent	169.7
125	25.5
150	0.13
175	<0.05
200	<0.05

(1) 60 minute hydraulic detention time

(2) EPA Method 625

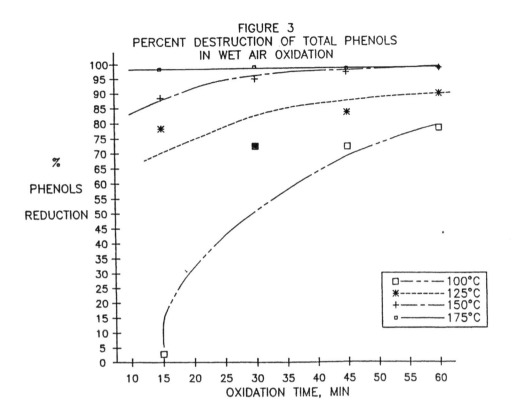

FIGURE 3
PERCENT DESTRUCTION OF TOTAL PHENOLS
IN WET AIR OXIDATION

The extent of oxidation of other organic constituents, e.g. hydrocarbons, that may be present in the spent caustic is dependent on the chemical structure of the specific organic compound. Some spent caustics contain various amounts of hydrocarbon oils. If the concentration of these oils is kept to a minimum, i.e. 100-500 mg/l, the evidence of oil is eliminated by wet air oxidation at approximately $200^{\circ}C$. If the concentration of oils is not limited and the oil concentration exceeds 1000 mg/l, then wet air oxidation temperatures of approximately $240^{\circ}C$ may be required to eliminate the presence of immiscible oils.

The residual Chemical Oxygen Demand (COD) of the wet air oxidized effluent, as a function of temperature and time, is shown in Figure 4. The overall destruction of COD is due to the oxidation of sulfides, mercpatans, and organics. The lower residual COD values at wet air oxidation temperatures of 150 and $175^{\circ}C$ is due to the complete oxidation of sulfides to sulfates as well as to a greater destruction of organic constituents.

The oxidation of the organic components in spent caustic, e.g. oils, phenols, and hydrocarbons, generally results in the formation of sodium carbonate (i.e. carbon dioxide) and water, plus low molecular weight carboxylic acid anions (i.e. acetates), which are biologically degradable.

WET AIR OXIDATION OF AN ETHYLENE SPENT CAUSTIC

Spent caustic that was discharged from the scrubbing system of an ethylene plant was oxidized in a wet air oxidation unit. The wet air oxidation unit was operated with a reactor temperature of $394^\circ F$ ($200^\circ C$) and a pressure of approximately 400 psig. The wet air oxidation unit processed twenty five (25) gallons per minute using a nominal hydraulic detention time of sixty (60) minutes. The chemical analyses of the influent and effluent streams are shown in Table 5. These analyses show that the inorganic sulfide concentration is reduced to a non-detectable concentration level by the wet air oxidation treatment. The mercaptans in the spent caustic were present at a concentration of 7040 mg/l and were reduced to less than 0.5 mg/l. The oils and polymers were present in the spent caustic at a concentration of 168 mg/l and were reduced to 7 mg/l in the oxidized effluent. The total phenols in the spent caustic were present at a concentration of 287 mg/l and were reduced to 5.3 mg/l in the oxidized effluent. The influent spent caustic was black in color and contained a small immiscible film of oil on the surface. The oxidized effluent was water clear in appearance and did not exhibit any evidence of immiscible oils.

TABLE V

WET AIR OXIDATION[1] OF SPENT CAUSTIC FROM ETHYLENE PRODUCTION		
	Influent	Effluent
COD, mg/l	33,084	846
COD Reduction, %	–	97.4
Na$_2$S, mg/l	1,650	ND
NaHS, mg/l	26,290	ND
Total Sulfide Sulfur, mg/l	15,718	ND
Sulfide Sulfur Reduction, %	–	100
Total Mercaptans, mg/l	7,040	<0.5
Mercaptan Reduction, %	–	>99.99
pH	13.57	12.4
Oils & Polymer, mg/l	168	7
Oils & Polymer Reduction, %	–	95.8
Total Phenols, mg/l	287	5.3
Phenols Reduction, %	–	98.2

(1) Wet Air Oxidation Conditions

Reaction Temperature, $^\circ F$	394
Reactor Pressure, PSIG	400
Waste Flow Rate, GPM	25
Nominal Reactor Residence Time, Min	60

ND = Non-detectable

WET AIR OXIDATION OF A REFINERY SPENT CAUSTIC

A spent caustic from a petroleum refinery source was oxidized in a continuous flow, wet air oxidation unit. The wet air oxidation unit was operated using a reactor temperature of 515°F (268°C) and a reactor pressure of 1610 psig. The wet air oxidation unit processed 5.3 gallons per minute of spent caustic using a nominal hydraulic residence time of 113 minutes. The chemical analyses of the wet air oxidation influent and effluent streams are shown in TABLE VI. These analyses show that the COD of the spent caustic was reduced from an influent concentration of 108,100 mg/l to a concentration of 11,600 mg/l. The total phenols were reduced from a concentration of 15,510 mg/l to 36 mg/l. The sulfide concentration was reduced from 3010 mg/l to <1.0 mg/l. The analysis for mercaptans was not performed. However, the influent spent caustic had an intense methyl mercaptan odor while the oxidized effluent had a mild hydrocarbon type odor. The influent spent caustic was dark brown in color and contained emulsified hydrocarbon oils. The oxidized effluent was transparent, with a light yellow color. There was no evidence of any emulsified oils in the oxidized effluent.

TABLE VI

WET AIR OXIDATION[1] OF PETROLEUM REFINERY SPENT CAUSTIC

	Influent	Effluent
COD, mg/l	108,100	11,600
COD Reduction, %	–	89.3
Total Phenols, mg/l	15,510	36
Total Phenol Reduction, %	–	99.8
Total Sulfur, mg/l	3,580	3,090
Sulfate Sulfur, mg/l	570	2,910
Sulfide Sulfur, mg/l	3,010(2)	<1.0
Total Solids, g/l	88.6	59.7
Total Ash, g/l	57.1	50.2
pH	13.0	8.3

(1) Wet Air Oxidation Conditions

Reaction Temperature, °F	515
Reactor Pressure, PSIG	1610
Waste Flow Rate, GPM	5.3
Nominal Reactor Residence Time, Min	113

(2) Determined by difference (Total Sulfur – Sulfate Sulfur)

CONCLUSIONS

The wet air oxidation process provides an effective means of disposing of spent caustic liquors which are produced in the petrochemical and petroleum refining industries. These wastes are classified as RCRA characteristic hazardous wastes (D003) because of their sulfide content. In the wet air oxidation process, the sulfides are oxidized to sulfates and lower the sulfide concentrations to non-detectable concentration levels. The organic constituents in the spent caustic liquors, e.g. phenols, oils, and polymers, are also lowered to low concentration levels.

REFERENCES

(1)   Federal Register, June 1, 1990 "Land Disposal Restrictions for Third Third Scheduled Wastes: Rule", pp. 22520-22720.

(2)   Dietrich, M.J., Randall, T.L., and Canney, P.J., "Wet Air Oxidation of Hazardous Organics in Wastewater", Environmental Progress, $\underline{4}$, No. 3, August (1985).

(3)   DeAngelo, D.J., and Wilhelmi, A.R., "Wet Air Oxidation of Spent Caustic Liquors", Chemical Engineering Progress, March 1983, pp. 68-73.